WRITING SCIENCE IN THE TWENTY-FIRST CENTURY

WRITING SCIENCE IN THE TWENTY-FIRST CENTURY

CHRISTOPHER THAISS AND
ASHLEY ROSE MEHLENBACHER

SECOND EDITION

broadview press

BROADVIEW PRESS
Peterborough, Ontario, Canada

Founded in 1985, Broadview Press is a fully independent academic publishing house owned by approximately twenty-five shareholders—almost all of whom are either Broadview employees or Broadview authors. Broadview is supported by a collaboration with Trent University, a liberal arts university located in Peterborough, Ontario—the city where Broadview was founded and continues to operate. Broadview is committed to environmentally responsible publishing and fair business practices.

Library and Archives Canada Cataloguing in Publication

Title: Writing science in the twenty-first century / Christopher Thaiss and Ashley Rose Mehlenbacher.
Other titles: Writing science in the 21st century
Names: Thaiss, Christopher, 1948- author. | Mehlenbacher, Ashley Rose, 1983- author
Description: Second edition. | Includes bibliographical references and index.
Identifiers: Canadiana (print) 20250255782 | Canadiana (ebook) 20250255790 |
ISBN 9781554816811 (softcover) | ISBN 9781770489721 (PDF) | ISBN 9781460408957 (EPUB)
Subjects: LCSH: Technical writing. | LCSH: Communication in science.
Classification: LCC T11 .T53 2025 | DDC 808.06/65—dc23

Broadview Press handles its own distribution in Canada and the United States:
PO Box 1243, Peterborough, Ontario K9J 7H5, Canada
555 Riverwalk Parkway, Tonawanda, NY 14150, USA
Tel: (705) 482-5915
email: customerservice@broadviewpress.com

Broadview Press books are imported and distributed in the United Kingdom and European Union by:
Gazelle Book Services Ltd.
White Cross Mills, Hightown, Lancaster, Lancashire, LA1 4XS
sales@gazellebookservices.co.uk

European Union – Responsible Person (for official use only):
eucomply OÜ
Pärnu mnt 139b14
11317 Tallinn, Estonia
hello@eucompliancepartner.com
+33757690241

Broadview Press acknowledges the financial support of the Government of Canada for our publishing activities.

Canadä

Edited by Michel Pharand

Book design by Chris Rowat Design

Broadview Press® is the registered trademark of Broadview Press Inc.

PRINTED IN THE UNITED STATES

1 2 3 4 5 6 7 8 9 10 25 26 27 28 29 30

CONTENTS

ACKNOWLEDGMENTS

Chris: Since the first edition appeared in 2019, many profound events scientifically, culturally, and politically have made a new edition necessary, and have inspired and shaped the book you have before you. Most importantly, collaborating on the design and emphases of this new edition, and on the themes and design of each chapter, with Ashley Rose Mehlenbacher, Professor and Canada Research Chair in Science, Health, and Technology Communication at the University of Waterloo, has made this a richer, more incisive, and timely book.

The thinking that has gone into this new edition over six years follows on the many years of research, teaching, and exploratory conversations that led to the first edition—and so I have many people to thank for their inspiration and generosity.

First, thank you to the students at George Mason University and the University of California, Davis, whose writings about their own research interests have informed me about many subjects I'd otherwise never have encountered and whose passion for those many endeavors has always given me great hope for the future. I especially thank the STEM majors at UC Davis whose writings in my course since 2008 have shown the analytical and generative power of a systematic, multi-modal rhetorical approach to STEM communication.

Among colleagues at George Mason who nurtured my fascination with writing across disciplines I thank especially Donald Gallehr, co-founder of the Faculty Writing Program in 1978, and the late Jan Cohn, who, as chair of English and visionary in general education, gave this young (at that time!) assistant professor the freedom to conduct research in what was then the new field of cross-disciplinary writing studies.

STEM colleagues at GMU who provided insights over the years into writing in their fields included engineer Murray Black, physicist James Trefil, psychologists James Sanford and Robert Smith, chemist Suzanne Slayden, biologists Christian Jones and Larry Rockwood, nursing scholars Jeanne Sorrell and Mary Silva, mathematicians Randy Gabel and Daniele Struppa, and most especially the late Stanly Zoltek, the mathematician who chaired our writing-across-the-curriculum committee for many years and whose teaching showed how students could be inspired to love math through writing to learn.

Writing Program colleagues Terry Myers Zawacki, Ruth Fischer, and James Henry stand out among those who were important to my own growth as a researcher and who built a professional and cross-disciplinary writing culture at Mason.

Since coming to UC Davis in 2006, I have been blessed with colleagues in the University Writing Program, the Writing Center, and across disciplines whose dedication, and creativity, to teaching writing in STEM fields has been a shining example and ongoing inspiration. These include, among many, faculty members Kenneth Andersen, Rebekka Andersen, Katie Arosteguy, Melissa Bender, Amy Clarke, the late Marlene Clarke, Aliki Dragona, Brad Henderson, Scott Herring, Marit Macarthur, the late James McElroy, Donald Meisenheimer, Dan Melzer, Heather Milton, Sarah Perrault, Brenda Rinard, Katherine Rodger, Wrye Sententia, William Sewall, Lisa Sperber, Victor Squitieri, John Stenzel, Teresa Walsh, Karma Waltonen, Carl Whithaus, Nate Williams, and John Yoder. For the same reasons, I also thank former PhD students and now faculty Aparna Sinha, Michal Reznizki, Kendon Kurzer, Mary Stewart, and Michele Zugnoni.

Recent conversations about AI in the teaching of academic writing with UC Davis researchers Marit Macarthur, Lisa Sperber, and Carl Whithaus have informed my thinking about this emphasis in the second edition of *Writing Science in the Twenty-First Century.*

I thank Annaliese Franz, Angelique Louie, and Tammy Hoyer for nurturing many young STEM researchers/writers through their leadership of the Undergraduate Research Center. I thank editors Eric Schroeder, Pamela Demory, Amy Clarke, and Gregory Miller for publishing young STEM researchers in our undergraduate annual, *Prized Writing.*

From 2016 to 2022, faculty, visiting scholars, postdocs, and PhD students affiliated with Davis's Robert Mondavi Institute for Wine and Food Science gave me unique insights into the symbiotic relationship between sciences and language through participation in their sensory descriptive analysis panels. I thank especially the late Anita Oberholster, as well as Hildegarde Heymann, Jean-Xavier Guinard, Susan Ebeler, Annegret Cantu, Raul Girardello, and Tyler Simons, among others, for training and practice in these methods, and for conversations about writing.

Many thanks to the outstanding team at Broadview for all they have done to make this book a success: Brett McLenithan, Tara Lowes, Michel Pharand, and Judith Earnshaw. The thoughtful comments by adopters of the first edition have also enriched this new edition.

Above all, I thank Jean Thaiss, whose love of good writing, faith in me, and enthusiastic support for this project have, once again, made it all a joy.

Ashley: First, and most importantly, I would like to thank Chris Thaiss for trusting me to join him in working on a revised edition of his excellent *Writing Science in the Twenty-First Century*. It has been an honor to work on the new edition with him, and I have learned a great deal working with such a thoughtful, careful scholar as Chris. The opportunity to work on this edition allowed me to put to work many years of teaching science writing and communication, which I was first able to do at North Carolina State University where I was a doctoral student and instructor of record for a third-year undergraduate course, Communication for Science and Research. I learned much from the students in those classes. Later, at the University of Waterloo, I was a member of a cross-disciplinary committee tasked with creating a course for the Faculty of Science in science communication, to be taught to first year students, and I have happily been able to teach that course, Communication in the Sciences, now and then. Students in the course have brought critical challenges and disciplinary norms to the conversation that helped me think through the challenges students today might face in writing and communicating science.

I also benefit from working as the Co-Director of the Trust in Research Undertaken in Science and Technology (TRuST) network, and in that role have had occasion to talk to many scientists and researchers, who all made me think more carefully and critically about communicating science.

I would like to second Chris's thanks to the Broadview team, including Brett McLenithan, Tara Lowes, Michel Pharand, and Judith Earnshaw. Brett initially reached out to me and introduced me to Chris. For that, I am extremely grateful as it was the beginning of this enriching collaboration.

This work was undertaken, in part, thanks to funding from the Canada Research Chairs Program. As a Canada Research Chair in Science, Health, and Technology Communication, the program allows me to conduct research that directly informs my thinking on science communication.

Finally, I am grateful to Brad Mehlenbacher who shares with me daily conversations about science communication, from the goings-on at the research university where we both work to the rich life of our garden.

INTRODUCTION

WRITING SCIENCE FOR NEW READERS, WITH NEW TECHNOLOGIES, IN NEW GENRES

Science must have clear, persuasive voices. At no time in history has it been both more important and more possible for scientists to be able to influence public policy locally, nationally, and around the world to earn trust with those who are impacted by science. At the same time, science must remain true to its time-honored mission to seek the truth and to resist any temptation to come to easy conclusions at the expense of the careful, often tedious, unglamorous collection of data and open-minded analysis.

Meanwhile, advances in information and communications technology have provided science communicators with an astonishing range of styles, languages, and tools for carrying out the work of science and for sharing that work worldwide.

This introduction will lay out the rhetorical focus of this book: its recurring emphasis on the scientist-writer's obligation to fit the content of writing (the what) to the needs of a great range of readers (the who). Today's readers of science, technology, engineering, and math (STEM) possess great differences in knowledge, but they want to understand scientific breakthroughs and processes—and to use them for practical purposes toward ensuring human and environmental survival and success. Increasingly, "Arts" is added to STEM (for STEAM) and this book can also help those in the social sciences and humanities become science communicators.

This introduction will also explain how twenty-first-century technologies have both broadened the audiences for science and brought about new genres for science writing not before possible. Among these recent advances is the effective redefinition of writing as no longer limited to the conventions of alphabetic and mathematical language, but now able to better use visual, aural, and kinetic capabilities for showing and explaining phenomena to communicate about science.

This book is for you, if

- You are a *practicing scientist* who needs to write one or more of the following documents: grant applications; project reports for supervisors or agencies; articles in research journals; laboratory or field research reports; summaries or reviews for non-specialist readers; posters and oral-visual presentations for meetings or conferences; blog posts or other online writings; create multimodal communications including videos, infographics, or other visual elements; opinion or descriptive pieces for more popular venues such as newspapers, magazines, and broadcast media; advocacy articles for legislatures or government committees.
- You are a *teacher of a STEM subject* who wants to help students deepen their knowledge or wants to help students read, write, and communicate better about their STEM topics or a communication teacher looking to teach students about communication in the sciences.
- You are a *graduate student* who is part of a team or lab that needs to write any of the above documents, or who teaches or assists undergraduate STEAM courses.
- You are a *science writer, technical writer, or non-fiction writer* with an interest in science topics and in writing about science in blogs, newspapers, magazines, books, and other non-specialist venues.
- You are an *undergraduate college* or *high school student* with a serious interest in a STEM subject and who wants to communicate well in school and with other people as interested as you are.
- You are anyone who loves science-relevant topics and who wants to communicate your interests, discoveries, and ideas to others—but may lack confidence in your writing and communication.

Writing Science in the Twenty-First Century presents 13 concise chapters (plus this introduction) that offer guidance and examples to help you succeed in a broad range of writing tasks and purposes, all pertinent to the work of science and other STEM fields. We emphasize the *twenty-first century* because writing about STEM topics is much more varied now than in the past and reaches perhaps a billion more people because of Internet and broadcast technologies.

Many of our chapters will also address how genAI can both help and hinder writers' growth as rhetorically smart and effective communicators.

A FOCUS ON SCIENCE, BUT RELEVANT TO ALL STEM

This book will take most of its examples and lessons from scientific fields, such as the life sciences, chemistry, physics, and geology, but some examples will be taken from mathematics and engineering, so those in all STEM fields should find the book relevant to them. The "Writing in Science" classes that our colleagues and we have taught over years have included students from a very wide range of STEM majors, including engineering disciplines, agricultural and environmental sciences, computer science, and mathematics, as well as fields such as chemistry, biology, and physics, not to mention the many majors that have developed as combinations of sciences and technologies.

Students also come from such majors as anthropology and psychology, which rely on methods, materials, and genres that have much in common with all other STEM majors. The principles, processes, and techniques this book describes have been useful to these students, and we hope you will find them useful, too, and perhaps even inspiring. Thus, when we refer to writing in science in this book, you should understand it in terms of this more-inclusive STEAM context.

Here is an overview of what you can find in each chapter of the book:

CHAPTER ONE: WRITING TO REACH READERS

This chapter illuminates the act of writing as a person's (or team's) striving to achieve effective communication with a wide range of people who want and need to understand the writer's message. We must first dispense with the idea that "the data and facts speak for themselves." The data and facts can only be as powerful as the writers and speakers who interpret and explain them. This is the important work of knowledge building and sharing, and it requires a serious commitment from both authors and audience.

By examining contemporary examples from both peer-reviewed and more popular science writing, this chapter shows how effective science writers craft their writing to relate to the interests of their readers. Specific tools and techniques will be described—including a **systematic method** of both reading and writing with rhetorical awareness. Indeed, we must be sophisticated readers to understand the scientific literature we engage and to broaden our communicative abilities.

CHAPTER TWO: BUILDING EXPERIENCE AND CONFIDENCE IN WRITING SCIENCE

How we become capable communicators may take many forms, but underlying all of them is a confidence in our voice. Tools to be explained will include **writing to learn** in science and other powerful techniques of learning and thinking for the student. The chapter will describe typical obstacles for student writers in STEM fields and how writers, often with teachers' help, can overcome them. This chapter considers how technologies—including generative artificial intelligence (genAI)—can help us think about writing and communication and be ethically used in brainstorming and editing, but also surveys the limitations of these **technologies** as we develop as writers and communicators. Indeed, the focus of the chapter is the cultivation of one's own voice, and only secondarily how technologies can help writers develop fluency and confidence to take on greater challenges. At its core, the chapter illustrates how the important diversity of perspectives, experiences, and communicative styles make scientific communications richer and more rhetorically sophisticated.

CHAPTER THREE: "WRITING" REDEFINED MULTIMODALLY

Using examples from recent digital and print sources, including peer-reviewed journals, this chapter describes how effective science communicators have augmented conventional letters and numbers with visual language (e.g., photographs), aural language (sound files), video, animation, hyperlinks, and generative artificial intelligence tools. Online tools have become indispensable for scientists, but the range of possibilities in digital rather than print forms means one must make many rhetorical considerations when crafting a message. This chapter begins a process of showing writers easy ways to integrate some of these tools into their writing, and how to do so with rhetorical effectiveness. This process will be further elaborated in Chapters Eight through Eleven.

CHAPTER FOUR: WRITING SCIENCE ETHICALLY

While ethical conduct in the pursuit of scientific advances will be noted from time to time throughout the book, this chapter pays particular attention to ethical conduct in ways particularly pertinent to science writing and overall communication. As scientists gain greater access to broader publics through the Internet and new genres, science writers contend more and more with how to resist the urge to overclaim the importance of their work. They contend with acknowledging the limitations of their methods in the face of (1) stringent competition for scarce funding, (2) opposition from those who have been historically harmed by, for example, scientific or medical institutions, (3) opposition from those, often hostile to science, who never hesitate to expound their points of view, with little regard for accuracy, and (4) misinformation.

Ubiquitous use of generative artificial intelligence technologies also raises ethical issues for researchers, both in their own work and in their assessment of others' work. In this chapter, we will explore these matters as well as accurate citation of sources, when and how often to cite, how fully to report methods, how to realistically measure the importance of results, and how to realistically make claims about the implications of one's research. Each of these concerns will be concisely explored in this chapter and advice given.

CHAPTER FIVE: WRITING THE RESEARCH ARTICLE, PART I—ABSTRACT, INTRODUCTION, METHODS AND MATERIALS

Because publication in peer-reviewed journals is still the gold standard for academic research scientists, two chapters consider the conventional introduction-methods-results-discussion (IMRD) research article and offer advice on making the most of the opportunities this form gives to write clearly, accurately, and persuasively. Looking at excerpts from influential articles in such journals as *Biological Conservation* and *Science*, this chapter shows how reader-oriented writing can both be persuasive and adhere to strict standards of precision. Chapter Five will focus on abstracts, introductions, and methods and materials as part of the IMRD form.

CHAPTER SIX: WRITING THE RESEARCH ARTICLE, PART II—RESULTS AND DISCUSSION

This chapter will continue the pattern established in Chapter Five, but will focus on the writing of results and discussions. It will consider new patterns in citation (e.g., links to sources) and graphic display that can make journal writing more effective in reaching readers, among other recent stylistic innovations brought about by online publication tools and capabilities. In addition, this chapter (and Chapter Five) will consider ways in which the different sections of the IMRD format can be tailored to the needs and characteristics of different types of readers.

CHAPTER SEVEN: WRITING THE RESEARCH REVIEW

For both working STEM professionals and STEM students, reviews of research are essential tools in both understanding scientific inquiry and communicating the importance of any path of research. Not only is the review of research an essential aspect of all IMRD articles (in the Introduction of the report), but the research review is also a major form in its own right in the peer-reviewed literature.

This chapter offers advice for making the research review a purposeful, persuasive, and reader-focused work that demonstrates the writer's ability to summarize, paraphrase, and quote other researchers accurately and forcefully. It calls for an awareness of the history of science and the ability to probe how ideas change in science. In this chapter, the review of research moves way beyond the summary of key studies to become a tool to shape the significance of research and forge paths for new research.

CHAPTER EIGHT: STEM JOURNALISM—WRITING, READING, AND CONNECTING WITH BROADER AUDIENCES

This chapter describes the huge growth in popular (non-specialist) science writing in print and online magazines, newspapers, websites, and social media. Examples from a wide range of print and online sources in this chapter illustrate ways in which STEM writers can learn to communicate with broader audiences and in ways attuned to diverse audiences and today's multimodal expectations and tools.

The chapter spells out some of the implications of broader global accessibility, and what these mean for science communicators. The most profound changes have been the global broadening of the audiences for science communication and the formation of new genres and styles enabled by tools and influences from social media.

Although the peer-reviewed IMRD article and the research review are established forms in traditional print publishing, they, too, have been influenced by the emergence of new technologies and their capabilities, as well as by the global reach of the Internet. The chapter will include tips on voice, genre, and organization, as well as writing vividly to attract and maintain readers.

CHAPTER NINE: SCIENCE ONLINE—NEW READERS, NEW VOICES, NEW TOOLS

All fields require science communication online to some degree. Communication of science online has become a common way whereby new ideas—and new voices—in science have been recognized. For example, consider science blogs, which can be stand-alone sites by individuals and research teams or interactive adjunct sites to established companies, journals, and agencies. New styles, including freer use of first-person point of view and multimodal capabilities, provide many ways to communicate science. This chapter offers insights and advice for making the most of these opportunities, especially through a consideration of audience, genre, and the affordances and constraints of the platforms used to communicate online. Challenges in these spaces, from bots and misinformation, and the difficulties of cultivating an audience in an attention-saturated environment are also discussed.

CHAPTER TEN: CREATING POSTERS AND INFOGRAPHICS

Poster presentations at academic conferences are a traditional way of displaying research in a highly visual, but condensed, format. Posters enable text, but they feature emphatic graphics, usually charts and diagrams, often colorfully. The poster's more versatile, online cousin, the *infographic*, is another genre used for communicating science that is highly visual but condensed. Whereas text is still the primary tool of the poster, the infographic emphasizes the visual, though it can still display a surprisingly large amount of text. Scientists' ability to create visual presentations is an integral part of their ability to communicate with audiences of all kinds. Creating a slide deck for conferences and meetings is common. This chapter offers advice and models not only for making the visual aspect of presentations stronger, but also to emphasize and strengthen the interactive relationship of the speaker and the audience. In this way, the chapter also explores how text, oral communication, and the interaction of different modalities can strengthen our communication of complex information.

CHAPTER ELEVEN: CREATING ORAL-VISUAL PRESENTATIONS

Scientists' ability to give oral-visual presentations is an integral part of their ability to communicate with audiences of all kinds. PowerPoint presentations are standard at conferences and meetings, as well as becoming more and more common as alternatives to straight textual presentations online. This chapter offers advice and models not only for making the visual aspect of presentations stronger, but also emphasizes and strengthens the interactive relationship of the speaker and the audience.

CHAPTER TWELVE: WRITING SCIENCE WITH STYLE AND STYLES

Although science writing is caricatured as "dry" (parched? drought-stricken?) and boring, this can rather signal a lack of care in writing and communication. Clear communication helps in clear thinking and argument, which is exactly why we cannot separate the work of science from science communication. Indeed, unclear and vague writing can cover up shoddy methods and thinking (which readers are too confused by to investigate!). Numerous examples of AI-generated text have illustrated exactly such problems, too.

This chapter digs into common stylistic problems with science prose—cluttered paragraphing, confusing organization, too much jargon, too many irrelevant numbers—and offers simple methods to make the science researcher's prose as interesting as the work and ideas themselves. Use of many technologies can aid in the editorial process, and this chapter considers where such tools are appropriate and where they might hinder one's efforts to obtain a clear, engaging voice in their communication of science.

CHAPTER THIRTEEN: EDITING SENTENCES

While a scientist takes great care to keep a worksite clear, clean, and ready for work, sentences in science prose often can't do their work because meaning is hidden by lazy writing. This chapter puts sentences in sharp focus and offers tips for editing syntax, voice, length, tone, and word choice to make meanings stand out. *Editing* of sentences is distinguished in this chapter from the crucial work of *revision*, by which writers modify their ideas and presentation throughout the process of writing to make it more effective.

WRITING RHETORICALLY: CENTRAL TO EFFECTIVE STEM COMMUNICATION

Throughout this text, we will use the terms "rhetoric" and "rhetorical" to discuss the process whereby researchers and other writers can successfully communicate their ideas to the readers important to them. Why is rhetoric so important to science? How can it be most usefully understood?

Let's begin with a thought experiment that we call "Confronting the Rhetorical Wall."

> Think about the scientific topics that you are most interested in. Think about the work you do and the work by others that you read. Now think about the people to whom you have spoken about these interests and about what you have read. How would you characterize these people? What do they have in common? How are they different? Are some of them easier for you to talk with than others? Why?
>
> Are there some people you avoid talking with about your work because (1) they have a different point of view about the subject than you do, or (2) explaining what you do and why you do it would be so hard for them to understand that you feel it's not worth the effort?
>
> Are you ever frustrated to realize that some of the people whom you **most need to convince about your work** are in one or both of those categories? Do you ever retreat back into your lab or field site to run another test or take more samples or make more observations because you don't want to think about those frustrations?

If so, then you have run up against a "rhetorical wall." Rhetoric is an ancient idea that more or less means the art and science of persuasion. The Greek philosopher Aristotle (2006) wrote about rhetoric more than 2,300 years ago, but even then the need for rhetoric was very ancient. Rhetoricians have been writing about non-violent persuasion ever since, as cultures have changed and technology has advanced.

Because rhetoric is so powerful, it is incredibly dangerous. The people who have been most successful in the world—whether for good or for evil—have tended to be good rhetoricians. But rhetoric does not require truth to be effective, merely the appearance of truth. Thus, for good reason, many who love science distrust, even hate, rhetoric for how it has been abused.

Because science normally proceeds in pursuit of truth, the idea of learning rhetoric might not seem necessary. However, most of us can readily think of examples of how difficult complex subjects can be to understand and why learning to communicate effectively is important. We can also think of cases that illustrate that science is no longer separate from society and, accordingly, science communication must address the civic sphere as well as the scientific one.

For example, almost all environmental scientists think it should be obvious that humans have caused climate change in the twentieth and twenty-first centuries, because test after test shows this to be so in numerous settings. Nevertheless, polls show that one-third of the people in the US refuse to accept the idea of human-caused climate change, and way fewer than half actually act in their daily lives as if this truth mattered.

Why is this difference between scientific truth and widespread belief and behavior so drastic? And why do scientists who wish to have some impact on public policy and public behavior need to understand this difference? In other words, why do scientists need to learn rhetoric?

WHY SCIENTISTS NEED TO LEARN RHETORIC

Part of the fear of rhetoric in the scientific community is that researchers will need to oversimplify the language of science to communicate with non-specialists. But rhetoric can help us understand how to transform our communications to address the audience's needs rather than trying to simply provide them with scientific knowledge or information for its own sake. Certainly, oversimplification is a danger when, for example, years-long processes of research and the intricacies of clinical trials are reduced by advertisers and mass media to dramatic claims of breakthroughs, such as played out with mRNA vaccines during the COVID-19 pandemic or "scare stories" about the supposed dangers of vaccines through the debunked alleged link between childhood immunizations and autism (a case that also reminds us of the importance of non-ableist communications, which here would not frame autism as a scare nor seek to "cure" autism). Rhetoric addresses all these issues and more in a rich set of scholarly traditions we here draw upon to more effectively communicate about complex information and knowledge.

Later, Chapter Four on "Writing Science Ethically" focuses on the temptations to scientists themselves to exaggerate the value of research or its applicability in order

- to secure funding from public or private sources, or
- to draw attention within the science community.

Whatever the reason for the oversimplification of science, the problem is not oversimplifying to a less educated audience as it is misleading audiences that could—and should—be trusted to understand a more cautious, less one-sided message. That trusting audiences can work is shown by the success of systematic nutrition labeling on food products over the past thirty years, which has drastically changed consumer—and producer—behavior. Systematic labeling has demonstrated the ability of consumers—when given access to diverse products—to make choices based on more complex information.

Chapters Eight through Ten will expand on this need to reach broader, non-specialist audiences with accurate information that is presented with respect for the intelligence and knowledge of diverse groups of readers.

Learning rhetoric so that one can reach different audiences should better be thought of as

1. Working to understand why different audiences need different information, and
2. Learning how to use different *language, media, and argument strategies to meet those different needs.*

WRITING AND SPEAKING WITH RESPECT FOR DIVERSE READERS

An attitude of respect for diverse audiences is more realistic than an attitude of contempt. Recognizing that *certain audiences have greater expertise* in some crucial areas than do the researchers will go a long way toward helping scientists succeed in reaching those audiences. For example, a key purpose of government scientific research agencies is to provide evidence to legislative and executive committees as they ponder proposals to change laws on such issues as health, water, energy, parks, roads, and public safety. Indeed, it's difficult to think of an area of government where scientific thinking and study are not important.

But scientists often lack the knowledge that lawmakers and staffers have of the greatly varying concerns of different groups of voters. Scientific data collection and analysis that meets priorities of the researchers is often doomed to be thought irrelevant—if the scientists-writers are not aware of the needs and concerns of the policy makers and their constituents.

For example, publishing in-house technical reports on an agency website meant for voters and legislators will most likely confuse and bore, unless those reports are carefully introduced and summarized in a way that addresses the concerns of the readers. Moreover, if even highly relevant research is presented only in the numerical and theoretical languages of the scientists, then the relevance of the research will never be seen and surely not used. It is tempting for scientists to criticize the inability of non-scientists to read scientific discourse and theory, but if budgets are cut for scientific research because of the lack of persuasive communication with the people who fund those budgets, then blame is poor compensation. Blame may feel good, but it won't pay the bills.

EFFECTIVE TOOLS FOR STEM RHETORICIANS

Fortunately, contemporary science journalism and communication offers us many, many examples of highly effective writing that bridges the gap between the languages of the lab (and the specialty journals that speak them) and the constituent groups responsible for bringing knowledge of real concerns and human situations to policy questions. Periodicals such as *National Geographic, Catalyst,* and *Scientific American* have different missions, but each reaches a broad audience of readers, most of whom are not science researchers but all of whom have varying levels of interest and knowledge in science topics. In addition, major newspapers such as the *New York Times* and *Washington Post* publish both online and print versions that include science-relevant articles on a daily basis.

Science journalism is a profession that has burgeoned in the past three decades, and which includes not only many skilled writers, but outstanding photographers and, more recently, digital designers. Chapters Eight through Eleven will particularly study methods used by these writers, designers, and editors.

Science communication is another field of research that studies a range of ways to communicate about science and challenges to doing so, and which empirically tests approaches. In science, too, we find that peer-reviewed research journals themselves are changing in response to access by larger online audiences—and by the awareness that science communication needs to become more rhetorically effective for science to succeed—and also to challenges within the disciplines such as the "replication crisis" that can be addressed through rhetorical means such as pre-registered studies. Chapters Five through Seven focus on this need to change and how it can happen.

In Chapter One, we will look more intensely at the process of reaching readers with your writing: scaling that rhetorical wall for audiences of different perspectives and kinds of knowledge.

CHAPTER ONE

WRITING TO REACH READERS

Topics in this chapter:

I. To Write STEM Well, Learn to Read Rhetorically
II. Six Categories of Rhetorical Analysis and Planning: A Systematic Method

I. TO WRITE STEM WELL, LEARN TO READ RHETORICALLY

Let's begin with an exercise.

On April 16, 2024, this headline and short "teaser" sentence appeared in the online *New York Times* (Anthes and Lee 2024):

Should We Change Species to Save Them?

When traditional conservation fails, science is using "assisted evolution" to give vulnerable wildlife a chance.

As you read the opening paragraphs of this article, think about how the language the writer uses

1. achieves the specific purposes of the article, and
2. helps the writer reach specific types of readers.

What do you think some purposes of this article might be? Who are some types of readers who would be interested in the article?

If you identify more than one purpose and more than one type of targeted reader, you're on the right track.

Here's how the article, by the *New York Times* writer Emily Anthes and photographer Chang W. Lee (2024), begins:

> For tens of millions of years, Australia has been a playground for evolution, and the land Down Under lays claim to some of the most remarkable creatures on Earth.
>
> It is the birthplace of songbirds, the land of egg-laying mammals and the world capital of pouch-bearing marsupials, a group that encompasses far more than just koalas and kangaroos. (Behold the bilby and the bettong!) Nearly half of the continent's birds and roughly 90 percent of its mammals, reptiles and frogs are found nowhere else on the planet.
>
> Australia has also become a case study in what happens when people push biodiversity to the brink. Habitat degradation, invasive species, infectious diseases and climate change have put many native animals in jeopardy and given Australia one of the worst rates of species loss in the world.
>
> In some cases, scientists say, the threats are so intractable that the only way to protect Australia's unique animals is to change them. Using a variety of techniques, including crossbreeding and gene editing, scientists are altering the genomes of vulnerable animals, hoping to arm them with the traits they need to survive.

In contrast, below is a summary of the abstract of an article that the *New York Times* piece links to with the phrase "an intervention known as genetic rescue" (Anthes and Lee 2024). This article (Pavlova et al. 2023) appeared in the journal *Biological Conservation*. Think about what purposes might be served by this article. What types of readers would be attracted to this article?

Compare the purposes you identified for the *New York Times* article with those you identify for this one in *Biological Conservation*. Compare intended readers, also. What are the differences?

Benefits of Genetic Rescue of a Critically Endangered Subspecies from Another Subspecies Outweigh Risks: Results of Captive Breeding Trials[1]

1 See Pavlova et al. (2023) for the full text of the background, methods, and results at https://www.sciencedirect.com/science/article/pii/S000632072300304X.

Highlights

- Gene flow from another subspecies is perceived as risky and often avoided.
- Such gene flow is needed to save critically endangered helmeted honeyeaters.
- Captive trials revealed gene flow between subspecies improved fitness.
- Gene flow between subspecies also resulted in fewer male-biased offspring.
- Releases of admixed birds into the wild should benefit population recovery.

After these "highlights," the abstract goes on to cover briefly the typical sections of the peer-reviewed experimental article (see Chapter Five): its introduction, materials and methods, results, and conclusions, in a single paragraph of about 250 words:

- Four sentences introduce the issues of "inbreeding depression, chance loss of beneficial variation, and reduced adaptability to changing environments" for species at risk of extinction, as well as the value of "genetic rescue via gene flow from genetically diverse sources." But this process is risky when "the only potential source of immigrants is a different subspecies that diverged long ago."
- In regard to methods and materials (four sentences), the "helmeted honeyeater *Lichenostomos melanops cassidix*" is described as the subject of these "captive genetic rescue trials," and the multi-part gene flow rescue process is summarized as employing the "neighboring subspecies *gippslandicus*."
- The results are summarized in one long sentence, with emphasis on successful breeding, "more nestlings per nest," and "less male-biased chick sex ratios."
- The concluding two sentences recommend "continued monitoring of fitness after releases into the wild," and judge that "with potentially riskier rescue increasingly becoming the only option for many populations," this study is an "encouraging test case." (Pavlova et al. 2023)

(See Chapters Five and Six, which cover the format, features, and rhetorical characteristics of peer-reviewed experimental journal articles.)

How do you account for the differences in content, style, and language between these excerpts of the two articles? What similarities, if any, do you see?

To complicate matters further, here is a paragraph from *A Guide to Climate Change for Kids*, published by NASA (2024) in freely downloadable form at https://climatekids.nasa.gov/kids-guide-to-climate-change/, plus a screenshot of page 2 of the guide:

> What Is the Difference between Weather and Climate?
>
> The main difference is time. Weather is only temporary. For example, a blizzard can turn into a flood after just a few warm spring days. Climate, on the other hand, is more than just a few warm or cool days. Climate describes the typical weather conditions in an entire region for a very long time—30 years or more.

Exercise 1A: Try to be as specific as you can be in defining the purposes, readers, and language choices of the three samples described. To help you in defining these differences and features, try writing your observations—and then re-reading what you've written to see if you have anything to add or change.

After you have jotted your list, read the following section.

SOME DIFFERENCES YOU MAY HAVE OBSERVED, AND WHY THEY ARE IMPORTANT

Analyzing any piece of writing in terms of the purposes and audiences (intended readers) that you perceived in it means that you are **reading rhetorically**. In other words, you are thinking about how and why the writers of the article (or any other type of writing) designed and shaped the piece to achieve certain goals and to influence certain types of readers.

Reading rhetorically also means that you are looking not only for information in what you read, but that you are also looking for why the piece of writing was written and who the writer or writers were hoping would read the article. If your goal as a writer in STEM subjects is to be taken seriously for your scientific work, then you'll need to think beyond just "conveying information" and do serious thinking about the purposes for your own writing (what you hope to achieve) and the readers whom you want to pay attention to your work and whom you'd like to influence. You'll need to become a rhetorical writer.

COMPARING THE ARTICLES IN THE *NEW YORK TIMES* AND IN *BIOLOGICAL CONSERVATION*

So, let's apply rhetorical reading to the excerpts that began the chapter. We'll start by looking at two of the excerpts. Look at the list of differences you have jotted down in Exercise 1A. You may have noted that the *New York Times* article and the *Biological Conservation* article are published in very different types of publications. The *New York Times* includes articles on many subjects (e.g., politics, sports, entertainment), most of them not focused on science, and is read by people who differ greatly in interests, backgrounds, and educational focuses.

In contrast, *Biological Conservation* publishes reports of experimental research in scientific inquiries pertinent to the journal's title. Its readers tend to be scientific researchers themselves or people who have a deep interest in the methods, results, and applications of experimental biological and environmental research. This basic difference between the two publications is an important *rhetorical* difference. It helps us understand which readers the writers are trying to reach, and, as we shall see, the writers' purposes.

Accordingly, you may have noted that some of the language in the *Biological Conservation* abstract would be familiar only to other researchers in the field, such as the Latin names of the species and terms such as "inbreeding depression" and "gene flow." If you were to read the methods and results of this article beyond the abstract, you would also see an intense use of mathematical language—which is required in a report of this kind and which the writers would expect fellow researchers to understand.

In contrast, the language of the *New York Times* excerpt is easily accessible to people without a specialized scientific research background. Its scientific terms, such as "habitat degradation," "invasive species," and "pouch-bearing marsupials" (which is exemplified by "koalas" and "kangaroos" in the excerpt) are widely known by people of different educational levels and language backgrounds. Moreover, the topics that are emphasized in the excerpt relate to political and even personal issues that are shared by people across many interests, ages, and backgrounds—for example, "infectious diseases" and "one of the worst rates of species loss in the world" (Anthes and Lee 2024).

Indeed, by looking at the language and topics in the *New York Times* excerpt, you may have concluded that the major purpose of the article is to alarm readers about the rate at which the popular notion of "the Land Down Under" is under threat of disappearing. This purpose of alarm, furthermore, is meant to reach a very **wide and multibackground readership** whose concerns may not be primarily about Australia, but about the economic, cultural, political, or even personal effects of the same forces of change in lands everywhere. Because the potential audience of the *New York Times* is so large and diverse, the article may reach and influence people who could, for example, donate (or increase donations) to environmental organizations or, more broadly, look for signs of similar distress in the regions where they live. Again, because the *New York Times* readership is so large and diverse, it will inevitably contain a large number of voters, who might be persuaded to favor candidates who support federal or local funding for projects like the animal rescues described in the article.

An article like that in *Biological Conservation* would never directly reach so large and diverse a readership as that of the *New York Times*. Therefore, by translating the heavily technical language of the *Biological Conservation* article into the more broadly accessible language of the *New York Times*, **the newspaper provides a necessary service** to the science described in the journal.

You may have also observed that the abstract of the *Biological Conservation* article also emphasizes the positive outcomes of the attempted rescue of helmeted honeyeaters, but is much more concerned with **proving to a skeptical scientific readership that this confidence is statistically justified**. Where the abstract modestly notes that the results support the conclusion that the successful rescue offers an "encouraging case study" (Pavlova et al. 2023), the article as a whole rigorously describes the methods and results that lead to this confidence—as all fellow researchers in the field would demand and expect.

Together, the two articles perform the vital task of bringing to huge, diverse audiences the results of new research in an emerging and controversial field—that of using genetic and other means to help species adapt to a rapidly changing climate. As we will see in other chapters of this text, both scientific journals and widely read popular sources are necessary for the scientific effort.

THE CHILDREN'S WEBSITE IN THE COMPARATIVE MIX?

We include the third document, the NASA website on climate change for children, parents, and teachers in this comparison because it offers a *sharp rhetorical contrast* to the other two pieces—a contrast that is becoming ever more common and influential in STEM communication. In your notes on this document, you may have recorded that the colorful, highly visual, and simple (but multi-layered) design of the website will allow it to reach readers who might not read either of the other articles. Although the site does not specifically concern the genetic research described in the other pieces, the website is pertinent to our discussion because it can play the (also) essential role of bringing multiple climate-related topics to the attention of readers—for example the parents of young children—who can help form widespread opinion on funding for research. Imagine, if you will, the impact that a site like this one can have on parents and teachers trying to help the children explore and understand the world around them.

PUBLICATION VENUE AS RHETORICAL CLUE

We also include this document in our comparison because its source, NASA, while exceptionally well known in space exploration, is rarely thought of as important in understanding climate here on Planet Earth. In *reading rhetorically*, as we've already seen, important clues about the purposes and audiences of what we read and write can come from our prior knowledge of where the document is published. If we aren't previously aware of NASA's long history in weather and climate research, then we'll need to browse the site further to get a clearer sense of the value of the venue and, more important, understand its philosophy and the reliability of its information.

What information is included in the site and the kids' guide that might lead you to trust its advice to parents and children? What more, if anything, would you hope that the site designers and managers would include? If you were to consider writing for a publication like climatekids.nasa.gov, what would you want the website to be sure to convey to the readers you wish to reach?

THE NEXT STEPS TOWARD RHETORICAL READING AND WRITING

Doing a more thorough rhetorical reading of the entire articles from which the excerpts were taken would lead to an even richer, more nuanced understanding of the purposes and types of readers of these articles. However, our purpose in leading you through this beginning analysis has been to model for you a method of analysis that can help you, as a writer of STEM materials, advance your awareness of language and your versatility in using it in constructing documents of various types. In the sections that follow, we'll describe a systematic way of reading rhetorically that you can apply to your own reading and writing.

II. SIX CATEGORIES OF RHETORICAL ANALYSIS AND PLANNING: A SYSTEMATIC METHOD

1. Purposes
2. Intended Readers
3. Order of Information
4. Types of Evidence
5. Style and Tone
6. Graphic Presentation—and the Changing Definition of Writing

Study the six terms above. Then try using them in analyzing any piece of STEM writing that you'd want to understand better. We've already had a bit of practice in applying the key terms "purposes" and "intended readers," but you'll find that the other four terms will help you systematically probe the rhetorical identity and power of what you read—and can help you design and write STEM documents that reach the readers you want to influence.

First, some definitions and key features of each of the six terms.

1. PURPOSES

Note that we use the plural "purposes" rather than the singular "purpose." It is rare for a writer to have only one purpose for a document, no matter how simple the piece of writing. Think about something you have recently written or are about to write. List the purposes you would like that document to achieve. Even a simple document—a short email, a social media post, notes on an article you have read—is likely to have multiple purposes.

If you are planning to write a more complex document—a lab report, for example, or a grant proposal—try listing some of the purposes you have for that writing. For the grant proposal, an obvious purpose is to *secure the funding for which you are applying*. But other purposes essential to getting that funding might be these:

- to show the funding committee that your team knows all the pertinent prior research,
- to demonstrate that your approach fills a pertinent gap in the research, or verifies (or disproves) an existing methodology,
- to argue convincingly that there are valuable applications of your study to important needs,
- to show that you have carefully prepared a budget that will use the funds for specific and necessary expenditures,
- and so on (think of more purposes that you might add for a project of your own).

Spending the time to list the multiple purposes of the documents you create can also pay off by helping you to organize the document so that these purposes get the emphasis they deserve. As this book explores different types of writing in its chapters, **you'll see how achieving the purposes of these writings will be reflected in how they are organized** into sections and paragraphs and in the use of appropriate headings and other tools to show emphasis.

In writing rhetorically, you begin by thinking about the various goals you want to achieve, then you build your first draft of the document toward achieving those goals. Of course, as you write you will probably become aware of other purposes that you'll emphasize when you revise your draft. But when you rhetorically read documents written by others, you begin by reading the (usually) published piece and then trying to understand the purposes that led to the document. By using both of these processes in your reading and writing, you'll become increasingly better able to make your writing achieve your purposes.

2. INTENDED READERS

Again, notice that we've used the plural "readers" instead of the singular "reader." As with purposes, it's rare—almost impossible, actually—for any piece of writing to be intended for only one reader. How so? Because every piece of writing is first meant to be read by the *writer* and only afterwards to be read by someone else or by many others. The only writings that are meant for one reader are those that are meant to be read only by the writer.

Indeed, one of the most basic and difficult tasks for all of us writers is to get out of our own heads long enough to actually imagine what another person expects from our writing and what those readers feel as they read what we have written. Consider for a moment how even the simplest conversation between strangers often includes misunderstanding by either or both parties. Messages need to be repeated and reworded, and we often have to use both facial expressions and hand gestures to make our meanings clear and convey the right tone. We even sometimes have this problem with people we know well.

Nevertheless, we persist in trying to communicate with others because we need to make ourselves understood in order to accomplish most things, especially in a world where increasingly we need to *collaborate* with others and *learn from them* across countries and languages.

For several centuries—and particularly in our own century—writing has been our preferred technology for this communication, despite its difficulty. The miracle is that humans have worked out methods and conventions so that much of this communication occurs very effectively. This miracle is especially vivid in science, which depends on effective communication *within teams*, which are often made up of researchers from diverse cultural and linguistic backgrounds, who may have limited opportunity to meet in person.

Scientists therefore communicate much of the time online and via articles and reports. Learning to understand and accommodate the needs and expectations of other readers and writers—including the members of your own research team—is therefore essential in the scientific enterprise. Reaching and influencing the readers beyond your team is obviously also essential to science.

Chapter Two focuses on a series of **tools** with which inexperienced writers can gradually break through their anxiety about reaching readers, steadily develop understanding of the needs of readers, and build confidence in their writing ability. Consistent practice and feedback from others are essential aspects of this process, as described in detail in Chapter Two.

"Intended" readers? For the purposes of this chapter, let us look briefly at the idea of intended readers. As with listing the purposes you want to achieve, **listing the types of readers you want to influence** with any writing you do can be very helpful in designing any document.

Often this task is not difficult, or at least it seems easy. If you write an email to a colleague about the date and time of your next meeting, you can list that one obvious reader. But we've also learned to be well aware that email is not a private network. Even if you explicitly *intend* only your colleague to read the document, you also know that that message might be forwarded to others and that the network managers and possibly many others could have access to it.

So, your awareness of these other *possible* readers implies an *indirect intent* on your part to reach them. This awareness may mean that your message will be more matter-of-fact and less personal than if you had asked your colleague about date and time in a face-to-face encounter. In this way, those other possible readers affect how you write your supposedly simple message.

For a more complex piece of writing, for example that grant proposal we considered under "Purposes," your intended audiences will be multiple (though still less complex than the audiences for a piece like that article in the *New York Times*). Besides yourself, these intended readers may include

- the other members of your team, each of whom may have a slightly different stake in the proposal,
- the reviewers who will judge the proposal, each of whom may have a different perspective on and knowledge of your area of inquiry,
- the funding agency that issued the call for proposals and authorized (if not also wrote) the instructions for applicants.

Fortunately, that first group of readers, the members of your team, can give you feedback on the drafts of the proposal you write and can, if they are willing, help you brainstorm and plan the proposal. When our students work in teams on research projects, they typically use tools such as Google Docs and Google Drive to provide feedback on drafts and to make their individual contributions to the writing of their team proposals. Experienced teams of writers use similar tools. Through this give-and-take process, team members learn much about their teammates as readers and writers.

The so-called blind review process favored by many granting agencies means that little can be known about the specific identities of the second group of readers, the proposal reviewers. However, proposal writers can usually assume that those reviewers have been chosen for their stature as researchers and their own experience as grant writers, even if they are not particularly familiar with the specific research described in your own proposal. This shared stature and experience means that proposal writers can usually assume that reviewers will expect to see proposals that achieve the kinds of purposes listed above (p. 31). Still, the uncertainty about the actual reviewers of your proposal means that funding decisions will inevitably be made in part on the basis of factors that cannot be foreseen by proposal writers.

For an article like that in the *New York Times*, which might reach millions of readers with greatly differing agendas, a writer's listing of a few types of interested readers can help keep the design process manageable. For the less experienced writer, listing the readers who seem to be targeted in a piece like that on the genetic intervention in Australian birds can help you understand the necessary complexity of such an article, and begin to help you think about how you need to tailor your language to the interests of different types of readers.

Scanning the first few paragraphs of that article (p. 24), you can probably identify several targeted groups of readers, including

- those interested in Australian animal species, especially in regard to species loss,
- everyone interested in the effects of climate change and environmental degradation anywhere,
- those interested in the political and economic effect of species loss anywhere,
- investors in genetics research,
- biological and environmental research scientists.

The first few paragraphs of an article such as this one are likely to give you good insight into the audiences important to the writers and editors of a mass circulation publication such as the *New York Times*. Note that concerns vital to each of the audiences listed here are addressed in the first four very short paragraphs.

Exercise 1B: Using the excerpt of the *New York Times* article, cite examples that match the interests of the five audiences in the bullet points above. Can you add other audiences to this list based on what you've read in the excerpt?

The "general public"? No such creature. Note, also, that we omitted an audience that might be named "the public" or "the general public." We do this deliberately. There is no such creature as the general public. Science writers should resist the urge to lump all readers who are not practicing science researchers into a single, vague category with this meaningless name.

When we ask our students to identify the target readers of the science articles they read from newspapers and popular news websites, often their first reaction is to say the "general public," by which they mean anyone who does not read peer-reviewed science journals or who is not doing lab science. This overly simple distinction between a relatively small "insider" group of researchers vs. the utterly massive "outside" group of everyone else is not just a problem for those in science and technology fields, but is pervasive across academic disciplines and most occupations. The sharp distinction is part of an "us" vs. "them" perspective that makes effective communication with most people difficult, if not impossible.

Marketing strategists in business learned long ago that successful communication (aka advertising) required subdividing the mass of potential customers into identifiable interest groups—and then tailoring communication to these groups with respect to—*and respect for*—their interests and perceived needs. How to make meaningful subdivisions, and then get the right message to the right customer has always been the challenge for marketers. It is no coincidence that one of the most obvious achievements of information technology has been to aggregate an individual's choices of sites viewed on the internet into a profile of that person's interests and perceived needs. Less sophisticated, but also pervasive, ways to subdivide a clientele include surveys, such as customer satisfaction surveys, that have become commonplace in many businesses.

As sophisticated as academics are in their research, many have been remarkably naïve in their approach to bringing their messages to readers and users beyond their own small research communities. Whereas some research scientists have been successful in translating research results into ideas that resound with broader audiences, most need to find non-researcher collaborators—technical writers, journalists, designers, marketers—who have their finger on the popular pulse and have the language to reach the **many diverse groups of users**.

A major purpose of this book—and especially of this chapter—is to help you *understand and respect* the differences among groups of readers and to address the perceived needs of these different groups in your writing.

Respect is a key principle for your audience and for colleagues. With respect to public engagement efforts, research has shown that there are costs for some researchers that we should be mindful of in our work. Women and other groups of people that have been marginalized are sometimes attacked for research and public engagement. One's positionality affects the reception to their work, even when they follow best practices for public engagement efforts, including tailoring to specific audiences, as we have suggested. Being aware of such issues, we can in our own writing and communications be mindful of and appropriately address those who may face sexism, gendered discrimination, racism, antisemitism, or other forms of bias. As professors, we will often talk about the importance of addressing our women-identified colleagues as "Professor" rather than by their personal name, because we know from research that they are more likely to encounter bias that undermines their credibility. When writing about science, and the scientists who conduct research, be mindful of how biases might shape how we write about people. For example, it would be inappropriate to talk about a woman's appearance when writing about her research. When we think about writing and communicating for different public audiences, our conventions might sometimes be more vernacular (tailored to specific preferences of different groups of readers), but the norms and values of science, with attention to ideas and research, should be the primary focus of our communicative efforts.

3. ORDER OF INFORMATION

How you organize your writing *should* reflect *how you want* your readers to read it. Never assume that anyone will start with the first word of your deathless prose and read it straight through, not missing a precious word. We tell our students that they shouldn't be fooled by the fact that we read their entire drafts. We do it because it's part of our job to give them feedback on their drafts and help them to improve. No other reader will treat their writing in this kind and constructive way. Every other reader will read around until they find something that interests them and then pretty much ignore the rest. Only if the writing continues to *hold their interest* will they keep reading.

Experienced writers have learned to take advantage of readers' shifting attention by **designing their work so that readers pay most attention to what the writers want them to attend to**. For example, look once again at the headline and opening short paragraphs of the *New York Times* article. The bold, large-font title catches the reader's eye. The reader then moves to the next closest information—the "teaser"—and gets the dramatic, intriguing phrase "assisted evolution" to state how science is saving species (Anthes and Lee 2024). The following paragraphs are all short—only one or two sentences each—and each gives basic information geared to the interests of large groups of readers: What is happening to amazing Australia? Why has conservation failed? What are scientists doing to save these iconic species?

By ordering the information in this way, the writer is betting that the reader, regardless of interest group, will hold on and read more. But even if, as is likely, the reader moves on to another article, the strategy will have pretty much ensured that the *writer's main message has gotten across.*

Is the strategy of the team that wrote the *Biological Conservation* article different? Well, not really. Again, take a look at the abstract. First, the title is dramatic: "Benefits of Genetic Rescue of a Critically Endangered Species... Outweigh Risks." It hits the needs and concerns of many biological and environmental researchers and the funders who support them. Then, since the form is that of an abstract, the following text is brief, only 250 words.

In fact, the traditional abstract is headed by a short list of bulleted "highlights": a sort of abstract of an abstract to drive the dramatic message home. Moreover, the abstract itself conforms to *what scientist readers expect*: background sentences, followed by methods and materials, followed by results, followed by conclusions and recommendations. (Chapters Five, Six, and Seven study in detail the conventional organization of the peer-reviewed journal article.)

The abstract, as a brief summary of what is to come, satisfies the reader's tendency to keep reading only as long as the article keeps the reader's interest. But even the least attentive reader will probably read far enough to get the main message in the conclusion: "[w]ith potentially riskier rescue increasingly becoming the only option for many populations," this study is an "encouraging test case" (Pavlova et al. 2023).

In both cases, then, the writers (and the editors of the publications) have "front-loaded" the information—and the point of view—that they wish to get across to the most readers, even those least attentive. If readers want to obtain greater detail about "assisted evolution" and these experimental processes, they will have to look beyond the abstract of the *Biological Conservation* article or the first few short paragraphs of the *New York Times* piece. The most important lesson for rhetorical writers in this analysis is that **order of information matters**, and experienced writers and editors know how to manipulate order for fullest effect. In many cases, achieving that fullest effect means putting the main message for your readers close to the beginning of the article, where readers are most likely to see it without searching.

Exercise 1C: The next time you read anything, keep track of the order in which you read it. Do you start with the first word in the title and then march straight through to the end, skipping nothing? Or do you read selectively, skipping around out of order, and even ignoring whole chunks of the piece? Why do you read as you do? Is this your usual reading pattern—or do features of the work itself lead you to read it—or ignore it—in a certain way? Finally, how could your practice as a reader teach you about how you'd like to write, or not write?

4. TYPES OF EVIDENCE

Another way to understand the purposes and intended readers of a document is by looking for the *types of evidence* that the writer uses to persuade intended readers. We noted earlier, for example, that one obvious sign that the *Biological Conservation* writers are trying to convince research scientists of the efficacy of the new procedure is their attention to **specific results** in their brief abstract of the Australian birds study. Moreover, the abstract includes the introductions to

- a highly detailed and descriptive **methods section**, and
- a statistics-intensive **results section**,

two types of evidence that would be required by an audience of researchers. In contrast, neither the *New York Times* article nor the very colorful infographic panels from the NASA website use statistics or devote space to a detailed description of methods or results. Both do include **references** or **links to sources** that are more research-intensive, but the documents themselves are meant for audiences that don't demand methods descriptions nor detailed statistical analyses.

Yes, some statistics are used in the full *New York Times* article, but these tend to be simply expressed and limited to those that help to make the writer's argument succinctly:

> Nearly half of the continent's birds and roughly 90 percent of its mammals, reptiles and frogs are found nowhere else on the planet.
>
> The most inbred birds left one-tenth as many offspring as the least inbred ones, and the females had life spans that were half as long, Dr. Sunnucks and his colleagues found.
>
> [B]y 1989, just 50 helmeted honeyeaters remained, clinging to a tiny sliver of swamp at the Yellingbo Nature Conservation Reserve.
>
> [I]n 2023, just 34 lowland possums remained. The first genetic rescue joey was born at Healesville Sanctuary last month.
>
> Dozens of hybrid honeyeaters have now been released into the wild. They seem to be faring well, but it is too soon to say whether they have a fitness advantage. (Anthes and Lee 2024)

On the other hand, the *New York Times* article heavily depends on **quotes from experts** who have been questioned by the *New York Times*. Twelve different authorities are quoted in this article, and each is identified by their relevance to genetic rescue research, whether in Australia or elsewhere. Moreover, when they are quoted, the authorities do not use heavily technical language, but speak in a *conversational and perhaps emotional tone* more accessible to non-researcher audiences:

> Without some kind of intervention, the helmeted honeyeater could be pulled into an "extinction vortex," said Alexandra Pavlova, an evolutionary ecologist at Monash. "It became clear that something new needs to be done."
>
> "There was a lot of angst among government agencies around doing it," said Andrew Weeks, an ecological geneticist at the University of

> Melbourne who began a genetic rescue of the endangered mountain pygmy possum in 2010. "It was only really the idea that the population was about to go extinct that I guess gave government agencies the nudge." (Anthes and Lee 2024)

Note that in listing the types of evidence described thus far,

- statistics and statistical analysis
- quotes from experts
- descriptions of methods
- references or links to sources

we have commented on their appropriateness to reader expectations and have avoided ranking them in terms of value. There is a tendency among those who favor a particular kind of evidence to look down on, or be suspicious of, those who prefer other types. The problem with that dismissive attitude is that communication with diverse readers becomes more difficult and may cause readers important to the writer to turn away. **By studying how different types of writing handle the evidence question, one can learn how to reach more readers.** We get into more detailed advice about using different forms of evidence in later chapters devoted to different types of writing.

Exercise 1D: In addition to the types of evidence listed above, some common types include

- physical descriptions of persons, places, objects, etc.
- narratives of events or processes
- analogies to similar phenomena

Read a STEM-related article of your choice from a large-circulation publication such as the *New York Times*. If possible, choose an article that, like the one on the bird rescues, cites an IMRD report in a peer-reviewed STEM journal, which you should also review. As you read both articles, look for and mark the types of evidence that the writers use to reach their points of view about the subject. Match the evidence you find with the types listed so far. Do you find other types of evidence besides those listed here? How would you define them?

5. STYLE AND TONE

Style

Our students, when reading rhetorically, like to distinguish between what they call a "formal" style and an "informal" one. **Formal** implies to them such traits as a more sophisticated (or technical) vocabulary, more complex sentence structure, and more concern with following rules for uniform design, such as in the structure known as IMRD for introduction, methods and materials, results, and discussion, which is expected in most peer-reviewed STEM journals. The papers they write for their classes they regard as "formal," and the audiences they address, usually their teachers, they assume expect formality in the writing.

An **informal** style means to our students one that appears more relaxed, more casual, more conversational, and less restricted by rigid structural requirements:

- The vocabulary is more common—accessible to more people.
- Sentences are shorter.
- Conversations can be part of informal writings.
- They show more experimentation with fonts, colors, and use of visual effects.
- The audiences the informal stylist addresses are closer to being peers of the writer and are comfortable with a more free-wheeling approach by a writer.

Of the three samples we've looked at in this chapter, the heavily infographic NASA children's guide would clearly be the most informal, while the *Biological Conservation* article is the most formal. The *New York Times* article is somewhere in the middle: the students would find formal elements such as a care to cite the views of numerous researchers and many links to research journal articles. They would find informal elements such as the conversational comments of Pavlova and Weeks cited above. They would also consider informal the many photographs of Australian places throughout the article. But they'd also notice that in its overall design the *New York Times* article is careful to front-load its major messages and then to use the rest of the piece to expand on the major points, with evidence. So, on the whole, they might say that the *New York Times* piece is more formal than informal.

So, **style**, like **order of information** and **types of evidence**, helps the rhetorical reader understand the intended audiences and how writers are attempting to achieve their purposes with those varied readers. Analyzing style will help you as a rhetorical writer in the same way, as you vary your own style in attempting to reach specific audiences in appropriate ways. In the chapters that follow on specific types, or genres, of writing, we'll return to the question of style—and what is appropriate for different purposes and intended readers.

Toward the close of the book, Chapter Twelve looks expressly at style tips that cut across genres.

Tone

Equally useful as an analytical and design tool is **tone**. Tone is a term borrowed from music and from painting to indicate the **emotional mood** that the writing creates in readers. Musicians strive for a particular mood as they play or sing; painters use color and shading to create particular moods. One of our students' favorite tone words is "serious," which they use to describe all the peer-reviewed journal articles they read for their projects. Serious, as in "this is important and carefully done; take it seriously." In contrast, they might see a piece like the NASA guide for children as less serious, because its style is much more informal—however, they'd also see that its colorful, unconventional style is a clever way to get into a very *serious topic*, climate change, and to reach audiences, such as children, who might not be reached in a more formal and more serious way.

Tone can be used by skillful writers in many ways, sometimes loud and dramatic, sometimes quiet and subtle, to inspire different emotions in readers. For example, in the *Biological Conservation* article, the title, "Benefits of Genetic Rescue of a Critically Endangered Species... Outweigh Risks," creates a *positive or hopeful mood* by using the terms "benefits" and "outweighs." "Positive" and "negative" are two more of our students' favorite tone words; they capture an overall mood that a piece of even quite formal writing can create in a reader. If you look again at the abstract of the *Biological Conservation* piece, you may notice how the writers manipulate tone in the abstract, as they move from the "hopeful" title to a more "somber" and serious first few sentences—describing "inbreeding depression, chance loss of beneficial variation, and reduced adaptability to changing environments" for species at risk of extinction, as well as a process that is risky when "the only potential source of immigrants is a different subspecies that diverged long ago." But the tone brightens considerably with the results sentence, which emphasizes successful breeding, "more nestlings per nest," and "less male-biased chick sex ratios" (Pavlova et al. 2023).

Because writers and editors of peer-reviewed scientific literature are so careful to build a serious mood, readers can miss how the language is often **subtly catching hold of our emotions and influencing how readers feel** as they read.

More Tone Words. To help you understand and use tone in your STEM writing, think of words that capture emotions and moods. Below are some of them. Notice that all of these tone terms have opposites. If you think of each pair as representing extremes of feeling, you can use the pairs to measure (or graph) *how* and *why* a piece of writing is creating feelings of a certain intensity, as well as how you might create moods of a certain intensity in your own writing.

Agitated________________Calm
Angry__________________Contented
Happy_________________Sad
Violent_________________Peaceful
Frightened______________Secure
Hopeful________________Hopeless
Ordered________________Chaotic
Successful______________Failed
Energized______________Bored
Critical (Attitude)_________Praising
Critical (Condition)_______Stable, Improving

Exercise 1E: Read a peer-reviewed journal article in a subject of interest to you. As you read, try to be aware of how the article influences your mood. Using the tone words mentioned in this chapter—or others that seem right to you—list the mood you are feeling as you read. What words in the piece particularly are affecting your mood? Then note if your mood changes as the article proceeds. What in the writing, do you think, is responsible for those changes?

Keep in mind that writers and editors of peer-reviewed articles in STEM usually strive for a serious tone and a formal style that are not overtly emotional. What you will find by doing this exercise are relatively subtle emotions that can change your mood positively or negatively, but perhaps not dramatically.

6. GRAPHIC PRESENTATION—AND THE CHANGING DEFINITION OF WRITING

Increasingly important in STEM writing are the multiple ways that **graphic presentation** can be manipulated by new **tech tools** to reach and influence readers. STEM writers have actually been concerned to present data and concepts graphically for centuries, from the drawings and diagrams of Leonardo da Vinci and Andreas Vesalius, through the ubiquitous use of tables, graphs, and other figures in professional journals and posters, to the contemporary explosion of graphical variety in online infographics and videos. Not to mention the publications, both print and online, that demonstrate photographic splendor in seemingly infinite forms. In addition, as technology has enabled us to easily manipulate sound within video, the audio should also be considered as part of "graphic" presentation.

If you find it a bit strange to think of the concept of "writing" as containing all of these artistic media and technological developments, consider that in contemporary communication all of them are more and more linked in creating a **multimedia environment** for the "reader," who is more and more a consumer and interpreter of the multisensory impressions that come through in STEM communication. This reader—actually reader, viewer, and listener—is being manipulated by these writer-designers to think and feel in certain ways about the phenomena presented. In short, we can no longer think of writing and reading as transactions with written words and numbers only, though these are still a vital part of communication, in STEM or otherwise.

In the chapters to follow on specific genres of writing (e.g., infographics, blogs, popular journalism, and peer-reviewed journals), the roles of graphic presentation will be further considered. Chapter Three, for example, focuses exclusively on the many multimodal technical tools that today's STEM writers can use.

For the purposes of the present chapter, we want you to be aware of a few aspects of graphic presentation that can begin to help you incorporate the multimodal into your rhetorical reading and writing.

How Graphic Presentation Affects Your Reading of Words—and How You Can Make It Work for You

Look yet again at the excerpts at the start of this chapter from the *New York Times* and *Biological Conservation*. This time notice paragraph length. How long are the paragraphs in the *New York Times* excerpt? If you have been told at some point in your education that a paragraph must be X number of sentences long, but never just one, then the *New York Times* writer and editor have violated that dictum. Why have they made that choice? (For more on this aspect of style, see Chapter Twelve.)

Now, look at paragraph length in the abstract of the *Biological Conservation* article.[2] Note that the abstract has been presented as one long paragraph, which in total presents the four standard sections of a peer-reviewed

2 The article can be viewed at https://www.sciencedirect.com/science/article/pii/S000632072300304X.

journal article (see Chapters Five and Six for more on this structure). Do you find this paragraph easy to read or difficult, because of the lack of paragraph breaks? Note also, however, that the abstract has been preceded by a bulleted list of very short sentences called "Highlights." Do you find these highlights easy or difficult to read? Why do you think the article writers and editors have made the choice to use both highlights and the abstract?

Our purpose for this exercise is to draw your attention to relatively commonplace uses of **visual graphic features** in printed or screen texts, and how they can subtly influence how you read and pay attention. We like to have students do what we call the "eye test," in which they hold a page of print at a distance and then say where their eyes go on the page. In every case, they report that their eyes focus on the shortest pieces of text that are surrounded by blank space—e.g., the shortest paragraphs—and on bolded words or on bulleted lists surrounded by blank space. Rhetorically, using blank space, short pieces of text, bullets, and boldface judiciously, so that they don't just clutter a page and confuse a reader, can help a writer direct a reader's attention to the writer's most important points. The *Biological Conservation* "highlights" and abstract, as well as the title, "teaser" sentence, and opening sentences of the *New York Times* article, are good examples of these practices working successfully. (Again, see more about these practices in Chapter Twelve.)

Exercise 1F: Take a piece of writing that you are currently developing—or one that you have recently written—and experiment with changing its visual graphic structure in ways similar to the basic moves described in this section on graphic presentation. Your goal is to help your intended readers *focus their attention on what you regard as the most important sections of your text.* Look for ways to use some of the following features:

- bolded words or phrases
- headings
- bulleted lists
- italics
- links to other websites
- important numbers (not just numbers for their own sake; too many numbers lose attention, not attract it)
- short paragraphs (or subdividing longer paragraphs)
- anything else you want to try

Hint: You may find that this exercise forces you to reconsider what you want to regard as your most important ideas in this piece, or at least rearrange your text to highlight what's most important. If so, then that's an added benefit of the exercise.

CONCLUDING, THEN LOOKING AHEAD

Using the **six categories of rhetorical analysis and planning** is a systematic way to help you become a rhetorically savvy reader and writer. At first, applying the six categories may feel difficult and confining to you. You may find it easier to start by applying the categories to things you read, before you try them in developing your own writing.

Nevertheless, as you gain practice, you'll get more comfortable with the system and you may even begin to internalize the categories so that they become part of your writing and reading processes, as has happened for many of our students.

A bigger issue for you may be that the system might make you feel less inclined to write—even though the purpose of the system is to demystify some aspects of writing and actually make it easier and even more enjoyable. If you are someone who already has some anxiety about writing and who does not enjoy it, then we urge you to read Chapter Two, which acknowledges lack of confidence in writing in STEM and which provides a number of exercises and (we hope) easy techniques to get you writing more comfortably and becoming a part of the STEM writing community.

CHAPTER TWO

BUILDING EXPERIENCE AND CONFIDENCE IN WRITING SCIENCE

Topics in this chapter:

I. From Fear to Confidence
II. The Many Modes of Science Writing
III. Writing as a Necessary Tool for All in Science
IV. Overcoming Obstacles for Science Writers in College
V. When Knowledge and Practice Seem Unconnected: What to Do?
VI. Overcoming Obstacle 2: Lack of Helpful Feedback from STEM Instructors
VII. Building Confidence as a Writer in English or Another Language
VIII. Resources for Students to Build Writing Proficiency
IX. AI Pros and Cons: Can It Be a Confidence Builder?

This chapter is addressed most directly to students (the "you" in this chapter), but its lessons apply to all those who, at whatever career stage, lack confidence in their writing. Even the most proficient professional writers sometimes fear how readers will respond to their work.

Indeed, professional writers routinely use the tools described in this chapter to keep their skills sharp and help them negotiate the treacherous waters of new genres and readers.

I. FROM FEAR TO CONFIDENCE

A college junior majoring in life sciences wrote:

> Throughout my academic career, I felt I could only be good at writing or good at math, but not both—my test scores always seemed to prove this. Early in my education, state tests assessed three main categories: reading, writing, and math. I received higher math test scores, and the same pattern followed me all the way through my ACT, SAT, and AP exams. Thus, I felt a great disjunction between my ability to write and my interest in science as I was growing up. Since I chose a career path in the sciences, I thought I would not have to write as much as professionals in other fields. I had assumed writing would not be a large part of my career. I was wrong. (Excerpt from a student response to an assignment in a university course in Writing in Science)

The college student quoted above is one of many in the STEM disciplines—Sciences, Technology, Engineering, and Mathematics—who come to college thinking that writing is not relevant or important in the subjects that most interest them. Too often, the only writing they have done in high school has been in English classes. There, the main writing they have done involves an analysis of literature, a type of writing very different from what they might read on STEM websites or in STEM-related magazines or STEM textbooks.

Of course, someone had to have written all those STEM materials, but if students haven't written in their STEM classes, then they likely won't see the connection between what they have read and what they themselves, as budding scientists, need to learn regarding how to write in their majors, in grad or professional school, and certainly in careers.

Moreover, if the writing you did as a student in that English class was not something you enjoyed, and if you did not score well on it, chances are that you may not have felt great about yourself as a writer. You may even continue to fear being asked to write in any circumstance. Many of the students we have taught in university science-writing courses have come into those classes afraid of the writing we might ask them to do. They think of themselves as poor or inept writers. If those students also are not native speakers of English or only proficient in a language different from the one in which they had to write, they may have had other unhappy experiences with writing in school, or at least feel unsure of themselves. So their confidence as science writers may be very low indeed.

A main purpose of this book is to help students—and all writers of science—build confidence in writing. More specifically, it is intended to make you more knowledgeable and confident as a writer in the sciences and other STEM fields. The goal is for you as a writer not only to be able to handle course assignments such as lab reports and research reviews, but also to branch out and write enjoyably for readers other than your teachers and fellow students.

A large hope we have for this book is that you will learn to think of writing as a varied, powerful—even enjoyable—tool that can help make you an even stronger thinker: a person whose writing is an expression of

- careful observation and analysis,
- creativity and imagination.

Our larger hope is that you will come to look forward to writing, first for yourself as your reader and then for other people.

Our even larger hope is that you can become someone who can help change the world through your scientific practice—and through your ability to help many people understand, appreciate, and especially *use* the results and tools of science to save our planet and improve life for succeeding generations. Here in the present day, **so much of science is lost or misunderstood** because those with scientific knowledge fail to communicate with those—everyone—who need to apply this knowledge for survival in a fragile environment.

II. THE MANY MODES OF SCIENCE WRITING

Whatever the disciplines that you most closely identify with, the present time could be considered a Golden Age of Writing in those fields, and in science generally. Why? Because science writing is all around us and in a breathtaking array of media and forms and languages. Gone are the days when scientific knowledge was the exclusive preserve of a tiny elite of men (and very few women) with the time and money and access to written records that could inform their observations of stars or of the mysteries of cells. Only a century and a half ago—a blink of an eye historically—very few had access to relevant knowledge to speculate on the inherited traits of plants or the mechanics of flight. Even today, racial, class, and language differences keep many people around the world from full access to these kinds of knowledge and opportunities to study. However, new technologies have helped break down these exclusions.

With the invention of mass printing technologies, then of film and radio, that lack of access began to change. Today, billions of people around the globe have access through the internet and with the aid of affordable technologies to the latest discoveries of scientists and technologists. No one on Earth is untouched by the products of scientific research—few there are whose lives do not depend on their ability to make important decisions regarding air, water, land, and what they can grow and eat. More than ever, people in every culture must choose from among an array of options and decide how to live from day to day, how to plan for the future, and how to think about how their actions and choices affect those around them.

III. WRITING AS A NECESSARY TOOL FOR ALL IN SCIENCE

And how do we as individuals make those difficult choices? In almost every instance, those choices depend on the tools of writing—words and pictures—words written or spoken or sung; pictures still or moving. Words in English, Chinese, languages around the globe, digital code, and on and on. Pictures on television, on billboards, on phone and tablet screens, in magazines. Words and pictures together—on labels, in instruction guides, in digital materials of all kinds—articles, blog posts, Facebook pages, Instagram and YouTube videos, government and industrial reports, advertising, televised news and opinion, printed and digital books. The sheer number of websites devoted to scientific topics or the products of research is staggering, all developed in the past 40 years.

If the number of individuals with access to forms of science-relevant writing is in the billions, the number of those writing science is in the millions. More importantly, the number of individuals with the potential to write—through access by smartphones—almost equals the number of those who read.

So how can it be that any student in a college or university in this century could imagine that writing is not an important, indeed basic, part of a science education? Nevertheless, many students do not appreciate the importance of writing in their science education, and few have the confidence to act on that appreciation. There are three main obstacles to this appreciation and confidence.

IV. OVERCOMING OBSTACLES FOR SCIENCE WRITERS IN COLLEGE

1. The biggest obstacle for science students as they learn to write is **lack of opportunity to write in their STEM classes.**
2. The second biggest obstacle is **lack of helpful feedback from STEM instructors.** (**Note**: Grades are not useful feedback.)

3. The third biggest obstacle is **students not using all those opportunities and media** for writing outside their classes about their scientific interests.

The great news is that if you can overcome Obstacle 3, then you're on your way to developing skill and confidence in science writing, even if 1 and 2 remain obstacles in the school, college, or university you attend.

OVERCOMING OBSTACLE 3

Practicing your writing about science wherever you can is the best way to become proficient and confident. Try out one or more of these simple tools:

1. Start you own informal notebook, journal, or blog.

Begin simply by setting up a paper notebook, a Word doc, or an online journal (aka blog, through wordpress.com or a similar platform). Since practice is your goal, you don't need to publish your work at first. Writing for yourself lets you calmly try out words and ideas, and you can be your own reader. No one is looking over your shoulder. You are free to use this space to write about any topic you wish for any purpose. A few examples:

- describe as closely as you can what you observe under the microscope or in the field
- list your research methods in detail so another researcher might copy you
- write about how you got interested in your favorite subject and about the scientists who inspire you
- write a very rough draft of a grant application letter or an internship letter
- using one of your favorite pieces of science writing as a model, try to imitate that style or approach on one of your favorite topics
- write a rough draft of a response to an article in one of the journals or science blogs you read
- go back to an earlier entry and rewrite as it suits you
- imagine the future and write about ways that your scientific field could influence that future
- and so forth: topics and purposes are endless.

This journal or blog is meant to be fun. Keep that in mind. If the writing exercise is keeping you feeling good about your writing and thinking, then it's working. Remember, this is a practice space.

But it's also much more than that. Your blog (or paper notebook/journal) captures you at a moment in your growth as a thinker and doer. While right now you may not think of yourself as having much of value to say to the science community, you'll come to appreciate the value of your data collection and your analysis to you and maybe others around you. Remember that no one has exactly your educational background or experience. What you write—if you write *honestly*—will be original because it's yours. And it will be valuable to you, and maybe to others, because no one will have exactly your perspective on what you are writing about.

For example, in the blog Chris writes about gardening, birding, and climate, he doesn't claim that his planting and nurturing of tomatoes, herbs, lemons, etc., is original or full of sage advice to other gardeners. Someday it might be, but he has an incredible amount to learn, experiment with, and write about before it might become that. Still, the writing has value because it records observations and methods at specific times and situations in a specific place, a record that would otherwise be lost.

Moreover, what also makes these blog entries valuable to him at this moment is their reflecting on why he grows these diverse plants and the characteristics of the places and creatures that he studies.

He can't know until he writes the blog and publishes it if these blog entries can be useful or valuable to others, or why. But he does know that if he does *not* write these entries, there will be no chance for him to remember what, how, or why he did these things, and certainly no chance for others to see if his thoughts and deeds have meaning for them.

2. Write crowd-source reviews on Yelp!, Amazon, Netflix, or any similar site or blog, whether science-related or not.

One of the coolest things about today's internet is that we are always being asked to write our opinions of things. And many thousands of people take advantage of these opportunities to practice making themselves heard in a low-pressure environment. Facebook or TikTok, for example, give us many chances to comment in writing on what our "friends" have written. So you might at least sometimes resist the urge just to click "Like" or an emoji, and instead write some of your own words. All these sites are great, also, because you have the chance to read what others have written. In your own response you can be as original as you want to be, or you can repeat what others have said, only making a few changes to reflect your own thinking. As with the blog or journal described above, consider the review space as practice.

Moreover, these social networking sites allow you to gain practice choosing (or creating) and posting photos, videos, and audio files to enhance your words. As you will see throughout this book, using visual and audio material with your words is an increasingly important part of what we broadly mean by "writing."

While social-networking sites like TikTok or Instagram are not particularly STEM-focused, you can choose sites and blogs that are (see Chapter Nine for more on STEM blogs and writing them). As you read one of your favorite STEM blogs, imagine crafting a response, or at least a detailed question, to the blog writer. Try it out in your practice space. When you become sufficiently confident to write that comment or question to the blogger, do so. You can tell by the lack of comments to most blogs that writers are actually hungry to get responses, so don't think that the writers will be offended or annoyed by what you write. Instead, remember that people publish their blogs because they want to be part of a conversation with other people who take their ideas and work as seriously as the bloggers do.

If you get a response to your comment or question, bravo! You've become part of the community of science writers, even if you're just beginning. Carl Sagan, Lisa Randall, or Neil deGrasse Tyson had to start somewhere. Even if you don't get a response right away, don't let that stop you. Practice is the most important purpose, and when you do get that response, you'll already have enough practice to make your next response even easier.

3. If you are a student, use opportunities on your college campus or at your high school to communicate your scientific interests.

At our universities, there are several student publications that give undergraduate STEM students forums to publish their interests. Here's an excerpt from a post on the local blog "One Health" by an undergraduate major in global disease biology:

> On a trip to Hendy Woods State Park in Mendocino County, Professor Foley, Risa and I trapped and processed deer mice (*Peromyscus maniculatus*), looking for the presence of infectious disease-causing bacteria. We collected blood, ticks and ear snips from each animal, all of which involve methods that are harmless.
>
> **Teamwork, initiative and communication were key**
>
> Our fieldwork required collaboration, initiative and effective communication. We all took turns doing different tasks to collect samples efficiently. I learned how to handle the animals precisely and safely, and take the correct steps in gathering data.
>
> I tend to have a soft voice. I didn't notice it could be a communication issue until my team constantly found themselves asking, "What?" whenever I spoke. Professor Foley made sure I understood

that speaking up in a strong voice will come in handy when I am a doctor helping a patient in chaotic situations. Being heard could save a patient's life. (Sanchez 2016)

OVERCOMING OBSTACLES 1 AND 2

Obstacle 1: Lack of Opportunity to Write in STEM Classes

If you are a student taking a very large introductory or even major-level course in a STEM subject, and if your performance in such a course is measured only by your taking multiple-choice or other machine-gradable tests, you will need to find your own ways to use the power of writing to help you learn the subject matter and think carefully about it. Remember: multiple-choice and similar tests are not intended to help you learn the subject, but only to show your teacher or teaching assistant that you have already learned that portion that shows up on the test.

Failure rates on such tests are high *not* because students aren't capable of learning, but because many who fail have not practiced the techniques—often requiring writing—to help them learn more and better. For example, one such tool—doing short pre-writing exercises—is the subject of an article in *Science* (Kizilcec et al. 2017) devoted to improved completion rates in Massive Open Online Courses (MOOCs).

Striving for "Engagement" and "Interaction." If you attend lectures and just listen to them, and if you only read the course textbook, you are hoping that somehow the lessons' words and pictures will "stick" in your brain so that you can call them up for the test questions. We know from many years of research by psychologists, education scholars, and writing specialists that this "sticking" method is not how people learn. How we do learn is through what is often called "engagement" and "interaction" with the ideas and terms we are trying to learn. What do "engagement" and "interaction" mean?

You can get a good sense of "engagement" and "interaction" by looking at how you have learned any subjects or skills that are important to you:

First, you spend a lot of time with them—perhaps more than some people in your life would want you to spend. You spend this time because you enjoy the activity. Often, the more you do it, the more you love it, even as the difficulty of its challenges grows.

Exercise 2A: Think about a subject or activity that you really enjoy, and all the ways that you practice or strive to improve your performance. Make a list of these ways:

- What have you read or studied to learn about and become better at this activity?
- How, how often, and in how many different ways do you practice the activity?
- Do you talk, correspond with, practice, or compete with others who have the same interest?
- Think of milestones of improvement you have reached? How do you mark those gains?

Second, you don't just passively absorb what others are saying to you or what you see on a page or screen. Yes, that passive absorption is part of learning, but if you want to learn, you go beyond. You ask questions about what you are hearing or seeing. You look for sources to help you answer the questions, and you seek out other people to help you answer those questions. In the act of formulating a question, you begin the process of using writing to further your learning, even if that question does not at first take the shape of a written question. Asking any question forces you to use your own language to understand what you have seen or heard—and any time you are using your own words to communicate, you are writing, even if those words are not yet put by you on a page or screen.

But if you use texting, say, or a Google box to state a question, you are definitely writing. Researchers call this type of writing "writing to learn," as its main purpose is to help the learner understand, make connections with other learning, and apply what has been studied.

Third, your desire to learn takes you beyond absorption or question-asking to try to turn your new learning into some form of **action**. Those actions can be as varied as those of the avid reader of stories who tries creating characters and plots. Or the avid video gamer who continually tries to raise a score and imagines new games. Or the music listener who buys a guitar and begins trying to play. Or the machine builder who thirsts for more knowledge to solve the technical problems that inevitably arise. Or the gardener who experiments with different levels of water or sun exposure.

Whether your desire is to replicate something you have seen or heard or to turn it into something new, you begin using many other parts of your brain, and usually use other parts of your body, to create that action. You may require tools to bring about that action, and so you acquire other kinds of knowledge so that you can adapt those tools to what you want to perform.

In short, the process of learning requires that you "engage" multiple parts of the brain and by extension other parts of your body in the process.

True learning means a close and constant connection between acquiring knowledge and putting it into practice.

A main reason why STEM students go into their subjects is that they have already felt the strong motivation to put knowledge into practice. Indeed, many of our students come from families where putting knowledge into practice is how they live: for example, in health professions, in farming or raising animals, in engineering and mechanics, in electronics, in athletic or other physical performance. The relationship between classroom learning and practice in these fields should be natural for these students, and frequently it is.

V. WHEN KNOWLEDGE AND PRACTICE SEEM UNCONNECTED: WHAT TO DO?

However, all too often, the prescribed introductory learning that students encounter in their college courses seems unrelated to the student's motivations, and so that tight connection between knowledge and practice that is necessary for learning is missing. This lack of connection is especially true when the lecture course is not accompanied by a lab component or discussion section. Without an application and practice element, the lecture material can seem only an endless succession of technical terms and formulas that serve little purpose besides needing to be memorized for a test.

So how can the student—and the teacher—overcome this apparent disconnect between classroom and practice? Purposeful, science-relevant "writing to learn" can be one solution.

Here are two common writing tactics used by students to engage with their classroom learning and deepen the knowledge/practice connection.

1. Do real note-taking (not mere recording or "highlighting").

Real note-taking (or as a colleague called it, "note-making") means a combination of (1) close listening and watching, (2) jotting of occasional comments to help your memory, and (3) at least 15 minutes spent after the lecture devoted to writing about what you feel are the most important take-away ideas from the lecture. This real note-taking takes real discipline, because our tendency is to move on to the next thing in our busy schedules and just hope that the information has "stuck." This note-making is more than a summary of what you heard and saw, though it can also be that. Note-making is a chance for you to answer these questions for yourself:

- What was important in what I heard and saw, and why?
- How can I apply this information to my interests in science?

Merely recording a lecture for later viewing can be a good strategy for enhancing the practice of note-making, but it isn't a substitute. Similarly, the common practice by readers of highlighting or underlining sections of a text is *not* a way to enhance memory, although it can be useful if it helps you focus on sections that you want to write about in authentic note-making. Treat the reading you do for classes as you treat the lectures: as opportunities to "write to learn" about the reading or lecture you have experienced.

If you have never tried this type of note-making, try it the next time you hear a lecture or read a section of a text (such as this one). Watch what happens to your thinking as you write.

2. Take part in online forums and discussion groups with interested classmates.

More and more teachers are using their schools' learning management systems (LMS) to set up online discussion groups for their students and give course credit for that important work of sharing and understanding. These teachers are putting into practice for their students the idea of the scientific community of engaged learners and researchers. Many students are finding or creating such online forums for themselves, through tools such as Google Hangout or Google Drive, which enable both written discussion and contributions to group projects such as articles and reports. When our students develop team projects for our classes, for example, most teams use a combination of the tools on the LMS and the Google tools to plan projects, subdivide workflow, and revise drafts of reports and presentations.

What happens in these forums or on these workflow tools that is so crucial to learning? Two main things:

- First, you are running up against different points of view and different ways of using language that take you out of your comfort zone and challenge you to see the world in new ways.
- Second, you are being challenged by others who share your interests to come up with responses that all members of the group can understand and that will make use of the talents of each person.

But even if you don't have such opportunities for learning sponsored by your teachers, you can create these opportunities on your own with classmates, as students have for generations in informal study groups and clubs. Ask any student to describe their most powerful learning experiences in high school or college. They will likely name the extracurricular activities they were most involved in, where they could really talk with and work with other students on projects of mutual interest. The great thing about the online tools available now is that these discussions can be carried out even when you are not all together—and they allow you to develop your writing practice and therefore skill.

If you keep in mind that the national and international research communities in your field are using many of these same tools to carry on global conversations, then you will see that by engaging in these writing communities while in school, you are gaining valuable experience for your lifetime of learning.

VI. OVERCOMING OBSTACLE 2: LACK OF HELPFUL FEEDBACK FROM STEM INSTRUCTORS (NOTE: GRADES ARE NOT USEFUL FEEDBACK)

Getting helpful commentary—feedback—on your writing is essential for your growth as a writer. Useful feedback can take two forms: indirect or direct. For example, if you post on Instagram or Facebook and get a number of "likes" on your posts, that's indirect feedback, because it's always unclear what these viewers are liking: is it your writing or your visuals, the topic you are posting about, or just the fact that you are their friend?

Similarly, when a teacher gives you a grade on something you've written, that's also indirect, because you don't know what aspects of your work are responsible for the grade. If, for example, you are used to getting high grades on your work, and then you get a lower grade, why? And more important, what can you do to improve? The grade can't tell you. The only advantage of indirect feedback is that, when it is positive, it gives you incentive to keep going. But when it is negative, it can have the effect of turning you away from an interest or a topic that could have proved very rewarding if only you had had a bit of positive feedback to give you hope.

However, if you let negative feedback get you down, realize that it *can* be an incentive to get what you need: direct feedback.

Direct feedback addresses in some detail what you have written and lets you know what's working and what isn't. For example, the Facebook friend who writes back "Cool photo!" is giving you more feedback than one who just clicks "Like." One who writes "Great story! I'd like to hear more about your weekend. Did you go to the new exhibit? What did you think?" is giving you direct cues about what to write next or more about.

Similarly, if your chem lab instructor accompanies a grade on your report with a rubric that shows you scored high on the accuracy of your calculations but low on your discussion of the results, you have more information than the grade alone.

However, the rubric is still not as helpful as it could be, because you don't know the flaws in your discussion and how you might address them the next time you write. Direct feedback would require at least a comment or two about the discussion: e.g., "I can't see how you got to this conclusion from this result. Show your thinking." Better yet, the instructor could suggest a time for you to ask questions about the lab and the report and for you both to discuss how this report—and subsequent reports—might be improved. Such direct feedback is an incentive to deep learning because it provides an opportunity to put new knowledge into practice.

VII. BUILDING CONFIDENCE AS A WRITER IN ENGLISH OR ANOTHER LANGUAGE

Many of the students we have taught over the years, including a high proportion of STEM majors, speak and write more than one language. For most of these students, English is not their first language; nevertheless, many of them have studied English in school in other countries, often for several years. Although these students can understand what they hear and read in English very well, and can write and speak so they are usually understood by native English speakers, they often lack confidence to write in English in their courses. They try to avoid courses with significant writing assignments.

The same applies to those who study in other places where the language of instruction differs from their own first language.

Their fear of writing English or this other unfamiliar language may have been intensified by teachers or testers who seem only to have seen the mistakes in grammar and word choice in their writing. Such teachers disregarded the strengths in the students' language use and, more important, the content of their writing. When the STEM majors in our classes write about their sense of themselves as poor writers, they usually describe such negative encounters with teachers.

However, most of our STEM students who speak and write English as a second (or third or fourth) language write successfully because they have had the perseverance to look past negative feedback and keep going. In many cases, they have been fortunate to encounter teachers, especially in STEM subjects, who are not stopped by their reading of mistakes in grammar or word choice, and who also give them praise for their growth in language. Moreover, these teachers give direct feedback on the content of the writing:

its data, analysis, and conclusions. This direct feedback empowers the student to continue striving to learn and shows the teacher's respect for what the student is trying to achieve.

Here is one of our STEM students describing one such encounter:

> One experience that has stuck with me was when I had to write a report for an internship alongside a graduate student. Since I had some experience writing scientific reports before, I did not think it would be too difficult; however, I came to realize the difficulty in writing a report that was not about a standard laboratory procedure, but that was meant to provide new results in an interesting way.
>
> The graduate student that I worked with was very understanding that this was my first official report, and the grad student helped me guide my knowledge into writing meaningful information that anyone else could understand. I felt that helping the graduate student with the report was just as meaningful and educational as all of the previous lab reports I had written for classes, because it had a purpose to inform others about the research that was done and was not just a report done for a grade." (Excerpt from a student response to an assignment in a university course in Writing in Science)

VIII. RESOURCES FOR STUDENTS TO BUILD WRITING PROFICIENCY

The best resources are your teachers and your fellow students and writers. But you need to know how you can use them most helpfully—and ethically. (Chapter Four focuses on writing science ethically, and one topic we take up there is avoiding plagiarism.) There is a big difference between (1) asking a fellow student to read your draft of a report in order to offer advice, and (2) asking or paying someone to write your paper for you—or using ChatGPT to do the research and writing. The first choice is an essential part of building your own proficiency, while the second actions do nothing to advance your skill. (See Section IX below on the pros and cons of using AI tools such as ChatGPT.)

In order to get meaningful, ethical help with your writing proficiency from teachers and fellow students, you need to know what to ask for and how to ask for it.

Getting direct feedback from teachers. If you are taking classes in writing, your teachers are expected to be available to answer your questions about your writing, as well as to give you direct feedback on your assignments. Since most writing classes in high school and college are not focused on science writing (though many more should be), you'll need to seek out your teachers during office hours or online to ask specifically for feedback and advice on science writing you may be doing for a STEM class. It never hurts to ask for feedback on your STEM writing from a writing teacher, and many writing teachers will be happy to provide feedback if they feel they can. However, you'll need to understand if the teacher does not have the time to give feedback on assignments from other classes.

If you do ask for this kind of advice, be prepared with questions that will help the teacher focus on what you are most concerned about. Never just ask, "Will you please read my draft and tell me what you think?" This question is too vague; instead, be more specific: for example, "The goal for this assignment is to ______________; do you think my draft does that? How do you think I could improve it?" Or "I really worry about how clearly I organize my argument in the conclusion. Do you think it is organized? How could I make the organization clearer?"

It is certainly better if you ask for direct feedback from the STEM teacher who gave you the assignment. Even if this teacher (or teaching assistant) has not offered this feedback opportunity explicitly, do not hesitate to approach them with a specific request. Remember: ask specific questions about your draft, in order to help the teacher save time and to focus on what you most need.

Write a short proposal for what you intend to write to an assignment. Be sure also to ask your STEM teacher or assistant for feedback as you begin the process of writing to the assignment. If the writing assignment itself seems unclear to you, ask for a clarification. Once you begin thinking how you might address the assignment, consider drafting a short proposal for your teacher or assistant to respond to. This proposal should outline how you intend to address the assignment and why you have chosen that approach. Getting direct feedback at this stage of the process can save you much anxiety later on.

Getting feedback from fellow students. You should show the same courtesy and efficiency in asking for feedback from fellow students as you would when asking for feedback from teachers. Be prepared with specific questions about your draft rather than simply handing the student a bunch of your writing and asking "What do you think of this?" Remember also that the fellow student, unlike the teacher, probably has little idea about what the teacher is expecting from your writing. So the questions you can ask your fellow student will be limited by this lack of knowledge.

Getting "group feedback" from your instructor. If the student you are asking is also facing the same assignment, the two of you (or perhaps a larger group of fellow students) may decide that a more efficient strategy would be for the entire group to schedule an appointment with the teacher or assistant to get direct feedback on a proposal or set of proposals. We often have had pairs or small groups of students from the same class come to office hours for this kind of group consultation. Not only does this strategy save time, but each student in the group gets the benefit of the instructor's responses to the other students.

Teacher-organized peer feedback. The courses in which the best writing and most improvement are likely to occur are those which the teacher has designed to give systematic direct feedback to students on the writing they have been assigned to do. Most writing classes these days, especially if enrollment is capped at twenty or so students, are set up to provide students with the direct feedback they need to develop confidence and proficiency.

Unfortunately, it is still rare for such classes to be set up for STEM subject matter and for the typical types of writing that occur in STEM classes. But more and more STEM teachers and teaching assistants are becoming educated about how to give their students opportunities for teacher and peer feedback, especially in colleges and schools that take seriously the idea of "writing in disciplines and across the curriculum."

Using Writing Centers and Student Success Centers. Most colleges and universities in the US and increasingly around the world offer tutoring of student writing through what are called "writing centers" or "writing tutoring services" or "student success centers" or the like. For STEM students, it is much more common for such centers to be available than it is for there to be explicit courses in science writing, although that number is also gradually increasing. These centers are sometimes staffed by professional tutors or qualified graduate students, though it is more common for them to be staffed by undergraduate "peer" tutors who have received some training in how to give direct feedback to students who seek their advice.

If your school has such a center, take advantage of this opportunity. Whether you consider yourself a good or poor writer, you can always benefit from getting advice from another writer, especially one with some training in response to writing. While such center services are not the same as the feedback you can get from your STEM teacher or assistant, having that resource available can be rewarding in its own way. Be sure to read the policies and procedures for the center at your school on its website, so you'll know how best to prepare in order to obtain optimum results from the tutors.

IX. AI PROS AND CONS: CAN IT BE A CONFIDENCE BUILDER?

Several chapters in this book address the impact of artificial intelligence (AI) tools on STEM and on writing about STEM because that impact is profound. Here we look specifically at AI tools as they affect the focus of this chapter: your growth and confidence as a writer.

When ChatGPT was rolled out in Fall 2022, it caused immediate anxiety among all educators, regardless of their subject areas or the level of students they taught. What did they fear? Many assumed that a great many of their students, especially those in most need of a teacher's help, would just type their assignments into the bot, get back in a few seconds a grammatically correct, professional-sounding report, and—voilà!—assignment done, ready to hand in and get that A. Immediately, almost every university created committees and task forces to deal with the coming catastrophe to the university's mandate to improve student proficiency and certify the quality of its graduates.

Academic departments sought to create assignments and processes that were "AI-proof": for example, through in-class, timed-writing assignments and daily quizzes that couldn't be sent to ChatGPT, in addition to course policies that explicitly forbade students from using these tools under penalty of automatic Fs (or worse) if this "cheating" were discovered. In every case, reactions like these undermined the teacher-student relationship based on mutual trust. This panicked attitude assumed that most students were out to cheat and were too lazy or had too many competing activities, and so were uninterested in doing the hard work of studying and writing.

In contrast to these fears and suspicions, every technique described in this chapter depends on that mutual trust between students and teachers. Every technique assumes that students are devoted to learning, improvement, and realizing their potential. Every technique assumes that teachers and tutors will be accessible to students and ready to devote time to the careful consideration of your writing and to your questions about assignments.

Fortunately, many teachers across disciplines and many universities took a very positive attitude toward ChatGPT and other AI tools. They recognized

- that AI was not a new thing, that it had been around for many years in such applications as GPS, Google searches, translation programs, computer coding, medical diagnoses, etc.;
- that it would in fact continue to be improved and adapted to many different applications, as has indeed happened; and

- that because business writers and STEM researchers were going to have access to ChatGPT—and be expected to use it—why not find ways to incorporate AI tools into student learning in collaboration with teachers rather than in opposition to them?

USING AI TO BUILD A WRITER'S CONFIDENCE

It is crucial to distinguish between ways to use AI to build your confidence as a writer/learner and ways whereby AI can actually undermine confidence.

First, consider how AI can undermine your confidence:

- Assume that whatever AI produces is better than what you could write. (*Instead*, keep in mind that AI makes mistakes and that writers using it need to fact-check whatever AI produces.)
- Assume that if you ask ChatGPT to edit your writing for grammatical correctness, it will be "correct" for all readers and all purposes. (*Instead*, read the AI response closely to see if its word choices are accurate for your message, and if your intended readers will find the writing interesting and appropriate in its tone. Use the six categories of rhetorical reading from Chapter One to help you critique the AI response. Then revise your writing so that it serves your rhetorical demands.)
- Assume that your instructor will think that you are cheating if you use ChatGPT, and that you need to pretend that the writing is all yours. Such pretending—lying—totally destroys a productive teacher-student relationship and makes you think less of yourself. (*Instead*, at the beginning of courses, find out from your instructors what their attitude is toward AI and discuss with them how AI might be part of a writing process in which you are fully active and invested.)

Second, consider the following ideas for using AI to build confidence (these are just suggestions):

- Give ChatGPT a prompt related to something you want to write about, see what it produces, and then use the six categories of rhetorical reading to evaluate its strengths and to help you revise it to make it more useful to you. For example: you're writing a birthday message to a friend and you use the ChatGPT response as the basis for a revision that makes the message personal from you rather than sounding like a mass-produced card.
- Give ChatGPT a prompt that asks for a chart or other visual related to a favorite research topic or process in your own research area. Using the six categories, *revise* your ChatGPT prompt so that the visual and

its wording are more focused on your audience, your specific purposes, and your desired tone.

- Give ChatGPT a prompt that asks for an opinion on a controversial question that you have thought a lot about. Then write your own analysis of the response that you think is more accurate and less biased.

In short, always think of yourself as being in control of your writing process—and that what you produce is really yours, something that you would be proud to present and to discuss with teachers and peers. Consider ChatGPT and other AI tools as helpers in your process, but only as tools to learn from and to use as you wish as part of your authentic writing.

CONCLUSION

In the next chapter, we'll consider the exciting ways that science writing has been redefined through new "multimodal" technologies. As a writer in the twenty-first century, you have the opportunity to discover new ways to apply the "writing to learn" and other techniques described in Chapter Two to the writing world created by these new tools.

CHAPTER THREE

"WRITING" REDEFINED MULTIMODALLY

Topics in this chapter:

I. Do We Call It Writing—or Something Else? Multimodal Design, Perhaps?
II. Words
III. Numbers and Mathematical Symbols
IV. Photographs
V. Multicolor Charts, Tables, and Graphs
VI. Links to Other Sources
VII. Drawings and Diagrams
VIII. Video and Audio
IX. STEM Communication Online

A good metaphor might be represented in language or in visual forms. Consider the series of metaphors that Samantha Yammine, a science communicator with a doctorate in neuroscience and cell biology, shared in a video on social media (Science Sam 2023). There, Science Sam tells us that our "immune cells are pretty like the sun" with pictures of the sun juxtaposed with an immune cell, that our "heart cells are pretty like contemporary art" with a colorful collection of abstract arts juxtaposed with heart cells, that our "brain cells are pretty like fall foliage" with beautiful amber and yellow leaves on fall foliage juxtaposed with brain cells, and that our "fat cells are pretty like the ocean" with images of water reflecting light and, you guessed it, juxtaposed with a fat cell (Science Sam 2023). The video opens with Science Sam holding a phone and seeming to record using a mirror (a selfie-style video).[1]

1 The video can be viewed at https://www.tiktok.com/@science.sam/video/7310331265553616134.

Reading rhetorically, what seems to be the purpose of these juxtaposed photos in the video and their effect on readers? Would a series of textual descriptions of cells have the same impact?

Note that the video also features music. It is embedded within a website or app, providing numerous ways to enter. The images in the series supplement the text in the video, creating more impact than the text itself through the visual impact of juxtaposing our inner cellular worlds and the world around us. The photo reinforces what the video's text tells us verbally: this is one **purpose** that an accompanying photo can serve.

However, it is not the only purpose. The photos also grab one's attention and affect a reader's emotions. Because photographs do affect human emotions to varying degrees, communicators choose photos to move emotions in ways they intend, sometimes very little, sometimes profoundly. In this example, the emotional qualities are highly important, and this is demonstrated through the textual description Science Sam provides alongside the video: "next time you look in the mirror, remember this microscopic reminder 🥹 #science #scienceart #microscope" (Science Sam 2023). This affirming message also points us to several other modalities alongside the video, the photos, and the music! We also see hashtags, which are links to categories in a social media space, and one of those hashtags also points us to the idea of science art, which is another way to communicate about science.

Thinking rhetorically, a range of multimedia tools, including

- photographs,
- multicolor charts, tables, graphs, etc.,
- links to other sources,
- drawings and diagrams,
- video and audio,

make STEM communication in the twenty-first century more complex and very different from earlier eras of STEM writing and communication. Using visual and audio tools—along with words and numbers—will be a major emphasis throughout most of the rest of this book. This chapter introduces *why, how,* and *to what effects* STEM writers and researchers can employ these new and evolving communication tools. As in Chapter One, we'll look at these tools and opportunities rhetorically, examining

- how they can help you achieve the purposes of your research and writing, and
- how they can help you reach the audiences you want to influence.

Chapters Eight through Eleven will use the concepts and methods introduced here for specific applications to such genres as posters, infographics, blogs, presentations, and science journalism.

I. DO WE CALL IT WRITING—OR SOMETHING ELSE? MULTIMODAL DESIGN, PERHAPS?

You may be more comfortable calling this re-defined "writing" by a different name, such as "document design" or "multimodal composing." More and more, writing scholars are using **design** to capture the many conceptual facets that go into creating documents, as well as the wider range of creative and technical skills needed to build these multimodal works.

Basically, the core of a **multimodal** design approach to STEM writing is to take into consideration the roles played in successful written communication by

- the words, sentences, paragraphs, and larger sections of the written text;
- the apparatus of tables, charts, equations, etc., that make up the numerical presentation; and
- the many tools of visual and audio presentation that scientists can use to help tell their STEM stories.

In any given work, the balance of these key elements will differ, depending on such factors as the policies of the venue where you are publishing the work—and, more importantly, the mix of tools and information that can convey your thinking most effectively and accurately to achieve your purposes for your intended readers. **In other words, use the mix of tools that can have the most meaningful rhetorical impact**.

For example, many print journals in STEM still expect articles to consist mainly or even exclusively of words and numerical apparatus (tables, charts, graphs, etc.). Chapters Five through Seven focus on two genres of the peer-reviewed article (the IMRD report and the research review), and these chapters will heavily emphasize the rhetoric of these more traditional genres. But even here, online publication is enabling greater use of twenty-first-century presentation tools, which we'll also explore in this chapter and in Chapters Five through Seven.

EMPHASIZING VISUAL IMPACT

STEM journalism, especially magazine journalism, has turned to emphasizing greater and greater visual impact, in part to reach a much larger and non-specialist readership. Magazines such as *National Geographic* and

Scientific American use many **infographics**: detailed maps, diagrams, and charts that blend concise textual explanations with color coding, photography, and drawings to present complex displays that convey a plethora of information. In addition to magazines, museum displays increasingly use infographic design to convey, for example,

- historical trends and timelines,
- flowcharts of processes,
- developmental progressions, and
- geophysical, astronomic, and cultural maps, etc.

Infographics (and its traditional cousin, the poster) are the focus of Chapter Ten.

As infographic software becomes more readily available and user friendly, these tools have come increasingly within the creative reach of STEM communicators. Again, whether you use such tools or stay grounded in more traditional words-and-numbers presentation, you should be guided most importantly by the rhetorical demands of purposes and intended readers.

Let's look at each of the types of tools noted above (see p. 67) and consider how they can be used as part of STEM **rhetoric** in the multimodal environment.

II. WORDS

Words, put together in sentences, paragraphs, and larger structures, remain the core of STEM communication—words are the core of this book in most of its chapters. What scientists believe about any discipline and area of inquiry is most often expressed in non-mathematical language, whether spoken or written. Yes, among themselves, for example the members of a research team, communication can occur significantly in mathematical language or exchanges of diagrams and data tables, etc. Nevertheless, verbal interpretation of diagrams and data and placing data in a logical context always require worded explanations. The conventional structure of a report on experiments in STEM, that is,

- introduction,
- methods and materials,
- results, and
- discussion/conclusion,

depends on non-mathematical language used carefully and with insight into the expectations of readers.

RHETORIC AND THE TECHNOLOGY OF LANGUAGE

The history of rhetoric is most often the history of how humans have developed the technology of language—spoken, written, or signed—to influence hearts and minds, and to change how we understand reality. Those who are recognized as the greatest scientists down through the ages have almost always been recognized as highly proficient writers—by which we mean able to touch the emotions, reason, and imaginations of others through language.

Among the **skills** these workers with words exhibit are the abilities to

- describe perceptions of the senses;
- narrate clearly and step by step the processes they set in action, observe happening, or speculate will happen;
- compare one phenomenon or process with another;
- apply something observed, read, or realized in one setting to a different setting or situation; and
- argue cogently for an interpretation of data, especially while weighing alternative points of view.

Just as important, the proficient scientist-communicator is an astute, careful observer of how other humans use language—so that the scientist can choose words that others will understand, relate to, and be influenced by. When we look closely in this book at sample articles for how they use language, we are carrying out this effort to understand and learn from other language users.

Similarly, when we listen closely to presenters, teachers, or fellow learners and students, we are building our own skill to be able to communicate with these audiences. When we take the time to listen to and read the words of those who are not scientists, but who are affected every day by the results of science (in other words, almost everyone), then we become gradually more able to converse with and write to them about what we do. But if we do not take this time, then our ability to communicate is stunted.

Regardless of the other expressive tools we'll describe in this chapter, keep in mind that STEM writing never gets too far away from written or spoken language. Words are almost always an explicit feature of infographics, videos, audio files, and captioned photos. The tables and figures in STEM articles are almost always accompanied by written explanations or applications that help us focus and apply the data shown or pictured.

One way to look at the tools of STEM writing is to see words as basic and indispensable, but not always as primary. In video, for example, the visual may be primary in capturing and impressing the viewer, but at some point, words will be necessary.

How we communicate words is ever-expanding and this is important for accessible STEM work. When we think of the range of assistive devices that can help us communicate with one another, we must expand how these can be included in scientific research, and how those often multimedia forms can include even more people in STEM research and work.

III. NUMBERS AND MATHEMATICAL SYMBOLS

Numerical data, mathematical operations, and **language** play a vital role in the rhetorical effectiveness of writing intended for scientist audiences in peer-reviewed STEM journals (see Chapters Five, Six, and Seven, which focus on these journals). STEM communicators customarily rely on such tools as

- numerical data reported in tables,
- statistical trends shown in charts and graphs,
- equations and algorithms, and
- statistical tests

to measure the validity, reliability, and significance of research. Rhetorically, scientist readers demand and expect mathematical and statistical argumentation (1) to inform the methods of research and (2) to support the conclusions that articles reach.

IN STEM JOURNALISM AND OTHER POPULAR WRITING

But, as we have already seen in our analysis of the sample articles in Chapter One, non-specialist readers also expect some use of numerical data as part of the reports of experimental research in newspapers, magazines, and other online and print sources intended for wider audiences. As we have seen, **statistics** are an important type of evidence in STEM-related writing of all kinds. In Chapters Eight through Eleven, one goal will be to show how statistics can be presented effectively in a range of genres not intended for scientific specialists.

PURPOSES FOR MATHEMATICAL LANGUAGE IN STEM WRITING

Looking at use of numbers and other mathematical features rhetorically requires careful thinking about the purposes for which writers use mathematical language in STEM writing. Statistical methods and results provide expected **evidence to support argumentation** in research articles. In journalistic reports of research, well-chosen statistics are a favored means of convincing widely different groups of readers. But there is a marked difference in expectations for numerical language between researcher-readers and most readers of the summaries of research that appear in newspapers and magazines. Researchers expect that a second purpose for mathematical reasoning and language in the IMRD report is to **demonstrate thoroughness and reliability in methodology**. For example, it is common to describe and exemplify key methodological algorithms and equations in methods sections.

The methods and materials section of the IMRD report in a peer-reviewed STEM journal has no parallel in the journalistic summaries of research. Instead, as described in Chapter One, non-specialist publications typically favor other signs of credibility, such as testimonials and comments by other scientists, by other kinds of authorities, and by those affected by the results of science. If research methods are described in non-specialist publications, it is usually done briefly—except in those cases where a methodological dispute has become newsworthy for a larger population.

A third purpose of numerical data and language in articles for STEM specialist readers is related to the second. Just as researcher-readers expect a detailed methods section that includes mathematical tools, so do they expect, stylistically, that the writer will show **familiarity with mathematical language** relevant to the study. Put simply, writers use a mathematical style because readers (and editors) of peer-reviewed STEM journals expect it.

Unfortunately, this expectation often leads STEM writers into excesses that work against rhetorical effectiveness. How often have you read journal articles that seem to drown the reader in tables of unanalyzed data points? Or that report long strings of calculations showing how, yes, the researchers were hard at work, but that don't seem relevant to the argument of the study and that therefore distract the reader? In Chapters Six and Seven, we'll look at some ways to present statistical/mathematical language that satisfies reader expectations for these types of evidence—and that keeps the reader focused on the main thrust of the article.

IV. PHOTOGRAPHS

Courtesy of Chris Thaiss

Why photography? Think of photography as one major element of the need for visualization in all STEM fields. Although the bulk of scientific publication has been in words or in combinations of words and mathematical language, scientists routinely use **visual tools**—including photography, drawing, and diagramming—to create maps, outlines, and pictures for themselves as they do their work. Why? Because visualizing, just like language, is an essential tool of thinking for many people. Visualizing helps people conceptualize and organize their thinking. It helps them understand relationships not only of objects but of ideas.

Words and numbers alone often cannot do the basic job of communicating the ideas and discoveries and results of science. Pictures—and, increasingly, video—accompany words and numbers to make scientific thinking understandable and credible to those whose lives focus on things other than scientific research. In other words, the great majority of people, all those who live with the results of science and whose opinions, votes, and money largely determine if scientific research will continue, also rely on visualization to shape their thinking, just as research scientists do. Photography today is an indispensable visual tool in that process of learning.

Visualizing, however, does not always mean looking at a picture or graphic. This is important to remember when using images or other graphics because they should be explained for someone who may be using assistive technologies such as screen readers or refreshable braille displays.

STEM WRITING—A BLEND OF THE VISUAL AND THE VERBAL

While it used to be cost prohibitive to include photography, and quality images were challenging to include due to size and space restrictions, communication of science is becoming a blend of the visual and the verbal through

- the digital publication of more and more STEM journals;
- cheap, high-resolution, color photography that is ubiquitous online; and
- high-quality cameras and editing software that are with us almost all the time in our hand-held devices.

If we don't consider the rhetorical effects of these mingled pictures and words as we produce more and more of this blend, we will stay stuck in a past paradigm and ignore where and how we live—how we play and do our work.

Today, many of us use images routinely as part of social sharing online. When we share an image to Instagram, we make several rhetorical decisions, first in what photographs we take, which ones we choose to share, and then through the platform-driven decision-making for cropping, applying filters, choosing hashtags, and the descriptions or text we craft alongside the image. Science communication also requires a careful and thoughtful selection of images.

Exercise 3A: Using a phone camera or other digital photographic tool, take a series of photos to capture something that you feel is important to convey to an intended viewer about your STEM work. For example, a piece of equipment, a stage in a process, a result, a research setting, a research subject, etc. Try to pick something you believe a photograph could show *better* to an intended reader than could be communicated in a description in words or numbers only—perhaps if that photo were combined with a written or mathematical explanation.

Before you take these photos, consider what in the photo you wish to emphasize to the reader, and try to plan your shots to highlight that emphasis. (You might write for yourself about this topic to help in your planning.)

After taking the photos, select which ones best convey what you want that intended reader to see. Ask yourself: "Are there things in the photo that might distract the reader from the focus I want?" "What could I do to make the most important feature(s) stand out?" Think about what else you consider when posting to your own social media account.

In everyday social media use, we know that how we frame images, the angle we take a photo from, the lighting, etc., all of these factors affect what we convey to our audience (followers, friends, etc.). Sometimes these selections can be highly misleading. During the COVID-19 pandemic, many people set up home offices and thought about their backgrounds, and some honest people shared what was going on around those carefully cultivated backgrounds—the mess of our everyday lives including laundry, children's toys, etc. You can read these backgrounds in different ways: we protect the privacy of our home life when work and school suddenly moved in, or we are not honest about our circumstances. It's probably a bit of both, and social media is also often cited as not representing our lived reality.

Misleading representations might be a part of social media, but they should not be part of science or science communication.

Look at the two photographs of strawberry plants that begin this section; what can you tell about the plant from each photograph? What would you need a different distance or perspective in order to show? If your likely viewer knows nothing or little about this species, how might the photos be misleading or insufficient?

If you have access to tools that help you edit your photos to help you achieve the emphasis you want, use them. But you may decide you need to take more photos to get the effects that you want.

Editing tools have conventionally been used carefully because the manipulation of images in science is taken very seriously, as trusting in a faithful representation—knowing there are selections made—is crucial. The website **Retraction Watch** lists many examples of papers being "retracted" —taken out of the scientific record—because of dishonest or concerning image manipulations.

AI AND IMAGES

You may know that artificial intelligence (AI) is also having an impact on images in science. Several images were so bad that they received media attention, as well as from scientists, but these generative technologies are being developed and corrected quickly. While outrageous examples might be easily spotted today, it is unlikely that this will be the case even by the time this book goes to press. The use of AI in science is a topic that is evolving, and it is too early to say with full certainty what the norms and practices will be at the time you are reading this book. There will be new tools, certainly. Scientists will have addressed some of the questions about if and when it is appropriate to use them in scientific research. There will be new ethical questions, and we address some of them in the next chapter on ethics. But it is incumbent upon you to research and understand appropriate and inappropriate use for your discipline, for your studies or work context, and for the journal or venue you are communicating through.

If you do include AI in your process of communicating science, you should consider its impact on the trustworthiness of your message, and also remember that you as author are ultimately accountable[2] for its content.

V. MULTICOLOR CHARTS, TABLES, AND GRAPHS

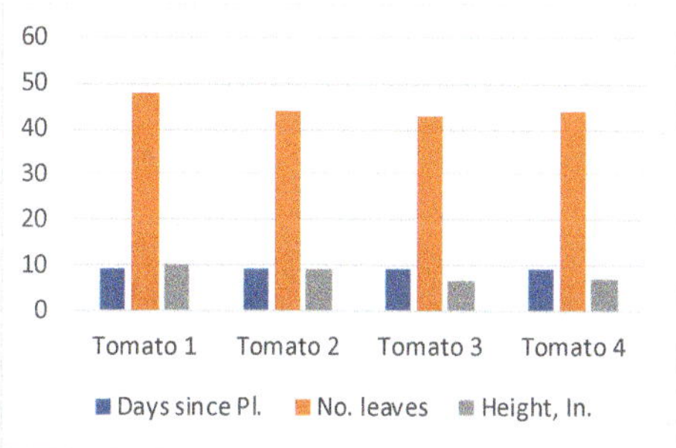

Tomato Growth (Set 1) Day 9

Ask yourself these questions:

1. What are the purposes of using charts, tables, and graphs to show numerical data?
2. How can you make your intended readers pay attention to these displays, rather than scanning them quickly (or not at all) in order to get to your worded explanation of relevance?

If your answer to Question 1 is "because research articles always have them," then your answer shows your awareness of the **stylistic expectation** of most readers of peer-reviewed journals. However, that answer is less rhetorically sophisticated than this answer: "because charts, tables, and graphs should help your reader see quickly why the numbers are relevant to your argument in the article."

Using these **display options** to present numerical data should be more meaningful to your intended readers than merely an unanalyzed set of numbers. For example, the simple bar chart above could be rendered as follows: "(1) 9 days, 48 leaves, 10 inches; (2) 9 days, 43 leaves, 9 inches; (3) 9 days, 42 leaves, 6.5 inches; (4) 9 days, 44 leaves, 7 inches." Even though the data set

2 For more about AI and questions of accountability in science, see https://www.lindau-nobel.org/blog-how-do-we-preserve-trust-in-science-in-the-age-of-ai/.

is very small, this linear display of the comparison of the four plants is hard to read, and certainly does not help the reader identify any significance.

In contrast, the multicolor bar chart provides a much quicker visual indicator of the similarities and differences of the four plants—at least in terms of the two factors (number of leaves and height) measured. Two items tend to stand out: (1) all four plants have between forty and fifty leaves each on Day 9; (2) the number of leaves for all four is similar, despite the fact that plants 3 and 4 are distinctly shorter than 1 and 2. The reader may be curious to see if these trends continue as the plants age.

Note how the data appear when a table is used:

Tomato plants, day 9	No. leaves	Height (in.)
1	48	10
2	43	9
3	42	6.5
4	44	7

Tomato Growth

Again, as in the bar chart, the **similarity** of the numbers of leaves stands out, as does the **difference** in height. In fact, you might feel that the table is clearer than the chart because the exact numbers are listed.

Now, note what happens when we add some colors:

Tomato plants, day 9	No. leaves	Height (in.)
1	48	10
2	43	9
3	42	6.5
4	44	7

Tomato Growth

The colors emphasize the sharpest difference in the data set, which the reader will expect the writer to comment on in the verbal explanation. Also, the color emphasis creates an expectation in the reader that this comparison of heights will be observed closely as the plants age. Building reader expectations is an important purpose for the use of visual displays of numerical values. Rhetorically, **expectation creates reader attention** and therefore ongoing interest.

Like charts, graphs are ideal for showing contrasts and similarities among comparable items. Graphs are particularly good at showing trends and progressions, because the line of a graph connects data points, whereas charts and tables do not. Therefore, they focus a reader's attention—if designed well. Note what happens when we extend the sample plant development through time for a single factor (height) and present it on a line graph:

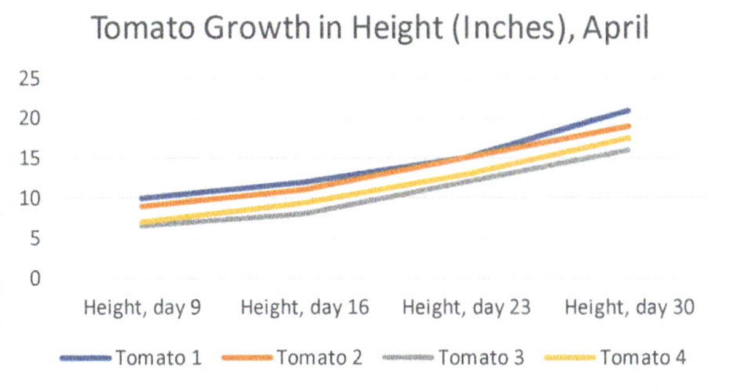

Tomato Growth in Height, April

Two results jump out at a reader: (1) the growth trend is very similar for all four plants; (2) the relative positions for all four have remained almost the same throughout the period. Note that the colors enable the reader to distinguish among the four at every point. Moreover, the line's trajectory creates reader expectation that the trend will continue, and so the reader will pay attention to how the growth pattern will play out over time.

DIRECTED ATTENTION VS. THE "DATA DUMP"

One decision the STEM writer always faces concerns the **amount** of data to display in the results section of an article. How can researchers balance the need for *thoroughness* in reporting relevant data against the rhetorical demand to *focus* the reader's attention on the most significant data?

In the sample graph, tables, and chart above, the ability of these displays to direct reader attention depends greatly on **restricting the number of categories and data points** that a reader will have to contend with. For example, if we wanted to add more categories and data points to the graph—number of leaves, stem thickness, and maximum plant diameter each week—we could do so, but we would have to consider what we might lose in the effectiveness of the graph in focusing the reader's attention.

Reject the data dump! Less experienced writers often choose to present far more data than are needed, perhaps because they lack the confidence to choose the findings that their research has indicated are significant. Part of this lack of confidence includes the fear that readers will think that the research has produced too little data to be significant—so the writer "dumps" it all on the reader, more or less saying, "Here, we did all this work—now you figure it out!"

Chapters Five, Six, and Seven will describe several ways whereby STEM writers can satisfy both

- the reader's desire to see that the research has been well-designed and thorough, and
- the writer's need to keep the reader's attention focused on the most significant findings.

Wisely using the contemporary options for numerical display via tables, charts, graphs, etc., will help meet both the reader's desires and the writer's needs.

VI. LINKS TO OTHER SOURCES

Digital publications have afforded new ways of moving through information. You can directly **link to a source**, unlike in print media. This changes the rhetorical function of a publication, not just its technical functions.

In the past, scientific publications were print-only. Citing sources from the literature of a field assumed that readers needed a full citation to track down a cited publication in a research library. In reading these citations, readers also had to assume (unless they already knew the source well) that the writer was accurately citing the source, and that this source was relevant to the inquiry in question. It became standard practice for writers to include long strings of citations (often in a single set of parentheses) and it was rare for readers to take the time to verify all the authorities the writer listed. This situation remains current in most print publications.

Moreover, unless the writer took care to list specific page numbers being cited (not just the author, journal, issue number, and date), it was rarely possible for readers to know why the source was being cited. Because, as we shall see in Chapters Four through Seven, it is not STEM convention in research journals to quote verbatim from cited articles, readers could not be sure what in the cited article was influencing the writer who had cited it. While it became commonplace for article reference lists to include 50 or more cited items, those lists came to function rhetorically more as a display of conscientious coverage of the literature by the writers than as a tool readers actually used to consult sources.

Today, print publications face the same limitation in terms of readers' ability to follow up on the citations in articles. Therefore, the standards for accurate citation remain largely the same as in the past (see Chapter Four for advice on proper citation). Of course, more and more of those citations are of online, web-accessible sources, and all citation systems (for example, Council of Science Editors [CSE] and American Psychological Association [APA]) have accommodated this newer type of source in their rules.

But online publications have different practices. Online publications are not limited by the same strictures that make consulting sources so painstaking in print books and articles. Online works can and should include links to other online materials. Of course, merely including a link does not mean that the reader will have access to the source, as many are paywalled. However, more and more sources are freely accessible and, if so, the writer can have the reasonable expectation that readers will consult at least some of them.

Similarly, link technology has made it possible and quite common for peer-reviewed articles to be set up like websites, with main and subpages available with clicks. Often, for example, the methods section of research articles, plus subpages of results and calculations, will not be part of the main article, but locatable via links. Scientists and researchers do not always read an article linearly, and links provide another tool (like subsections) to help readers get where they need to go.

LINKING VS. QUOTING

Quoting text in an article means that your reader is staying put. They are reading your content, and you are focused on the message that you want to share. Including a link can have a different effect. Think how often you have followed a link in one piece to another article, and then followed link by link from source to source—and never got back to the original article—and you may have forgotten what you were originally intending to do. So, the writer who includes links to other sources is rhetorically taking the risk that the intended reader will be distracted by the linked material.

Nevertheless, including the links can have several positive functions:

- First, linking to an authoritative source, such as the article in *Biological Conservation* (Pavlova et al. 2023) that was the source study for the *New York Times* article (Anthes and Lee 2024), adds to the **credibility** of the article that includes the link. Credibility is an essential rhetorical attribute, as it enhances what the classical rhetorician Aristotle called the *ethos* of the writer. Ethos can be roughly translated as "character" or "prestige" or "seriousness." Readers pay attention to ethos.

- Second, links can add **depth** to an article, in the same way as writers in print books add footnotes (or endnotes) that deepen the significance of portions of the text. Footnotes, which in most print books are printed in smaller font, take up less space than the main text and are considered by the writer to be less important than the main text, but nonetheless interesting and sometimes fascinating to the reader. Footnotes are often used to add citations to other sources. Many readers ignore the footnotes because of lack of time, and that is part of the invitation that footnotes, like links, offer. Links and footnotes say to the reader, "If you're interested and have time, here is something else you might find useful."
- Third, links can play the vital role in websites of being the lines of communication to other parts of the site. Links can create a **network of relationships** within the site, and they act as signposts to readers to guide them through the various pages of the site. Unlike a print text, which must be read linearly across a two-dimensional space, the network of links creates a virtual three-dimensional environment, somewhat akin to the network of neurons and synapses that connect three-dimensionally in the brain.

Websites that merely contain digital versions of print texts (as many do) serve an important purpose in the transmission of learning in all subjects, as they have made these works accessible to potentially millions of readers who could not otherwise benefit from them. But if you design your website to benefit from link technology to reach a range of other sources, you greatly increase that connectivity for those millions of readers, as well as deepen the teaching capability of the print works.

Exercise 3B: Select an article of interest to you from either a peer-reviewed online journal or an online science magazine. Mark all the links it contains (consider as a "link" any text or photo that you can click on to take you elsewhere on the site or to another source). Categorize each link according to the three positive functions described above. Note that a link may serve more than one of these functions.

What do you discover about how this writer has used links? Can you find in the article other opportunities for using links that might have further enhanced the article? Would you recommend that any of the links in the article be removed or changed? Why?

VII. DRAWINGS AND DIAGRAMS

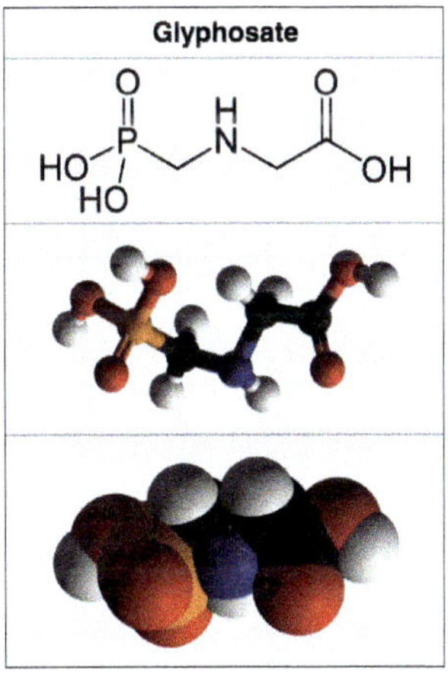

In STEM communication, drawings and diagrams have always played a key role in helping readers visualize what is being described or imagined verbally. These figures have always been essential in such fields as geometry and in the representation of elemental structures in chemistry.[3]

Before the advent of photography as an element in print publication, drawing and diagramming were indispensable to visualization. Think, for example, of the anatomical figures in the works of Leonardo da Vinci and Andreas Vesalius. Even after photographic images in print publications became more common in the late nineteenth century, drawing and diagramming were needed in science to visualize

1. phenomena that could not be photographed (e.g., cross sections of anatomical structures, atoms and molecules, astronomical charts);
2. phenomena that could be photographed, but that could be more clearly shown in drawings (e.g., plant root structures, architectural and engineering structures); and
3. theoretical projections and relationships (e.g., evolutionary trees, quantum mechanics, designs for flying machines, the inside of Earth).

Today, drawing and diagramming remain indispensable to helping readers visualize concepts in all three of the above categories. While advances in photography and editing have greatly increased the number and kinds of images that can be photographed and represented photographically, drawings and diagrams allow writers to

- focus the reader's attention on details of an object that photography doesn't highlight,
- show relationships among images (photographic or drawn), and
- allow verbal commentary in the diagram (e.g., infographics, such as the one pictured below).

3 See, for example, John Dalton's early nineteenth-century representations of atomic structures, as reprinted in Weber (2000, 45).

Drawing and diagramming, like photography, appear in STEM communication of many kinds, including journal articles, newspaper and magazine journalism, posters and infographics (shown at right), presentation slides, brochures, and many other forms.

"If This Tree Could Talk," UC Davis Arboretum, 2019. Photo courtesy of Chris Thaiss

As an example of multiple uses of drawings and diagrams in recent STEM publishing, let's look rhetorically at this screenshot of a news feature about an article on clustered regularly interspaced short palindromic repeats, or CRISPR, by Heidi Ledford (2016), in the journal *Nature*, published on March 7, 2016.

CRISPR: gene editing is just the beginning

The real power of the biological tool lies in exploring how genomes work.

Heidi Ledford

07 March 2016

PDF | Rights & Permissions

Illustration by Ryan Snook

Molecular biologists are riding a wave of new technologies made possible by CRISPR.

Whenever a paper about CRISPR–Cas9 hits the press, the staff at Addgene quickly find out. The non-profit company is where study authors often deposit molecular tools that they used in their work, and where other scientists immediately turn to get them. It is also where other scientists immediately turn to get their hands on these reagents. "We get calls within minutes of a hot paper publishing," says Joanne Kamens, executive director of the company in Cambridge, Massachusetts.

The theme of the article is embodied in the title and especially the subtitle: "The real power of the biological tool lies in exploring how genomes work" (Ledford 2016). The two most prominent visuals are

- The cartoon that heads the article, captioned "Molecular biologists are riding a wave of new technologies made possible by CRISPR"; this light-hearted drawing of lab-coated scientists on surfboards in a huge blue wave dominates the start of the article.
- The large infographic titled "Hacking CRISPR" (Ledford 2016) which includes large colorful diagrams of the five different CRISPR applications described in the article. The diagrams feature DNA being manipulated in different ways, usually involving "scissors," which have come to be the best-known image for the metaphorical "cutting" done by the Cas9 enzyme vital to the CRISPR process.

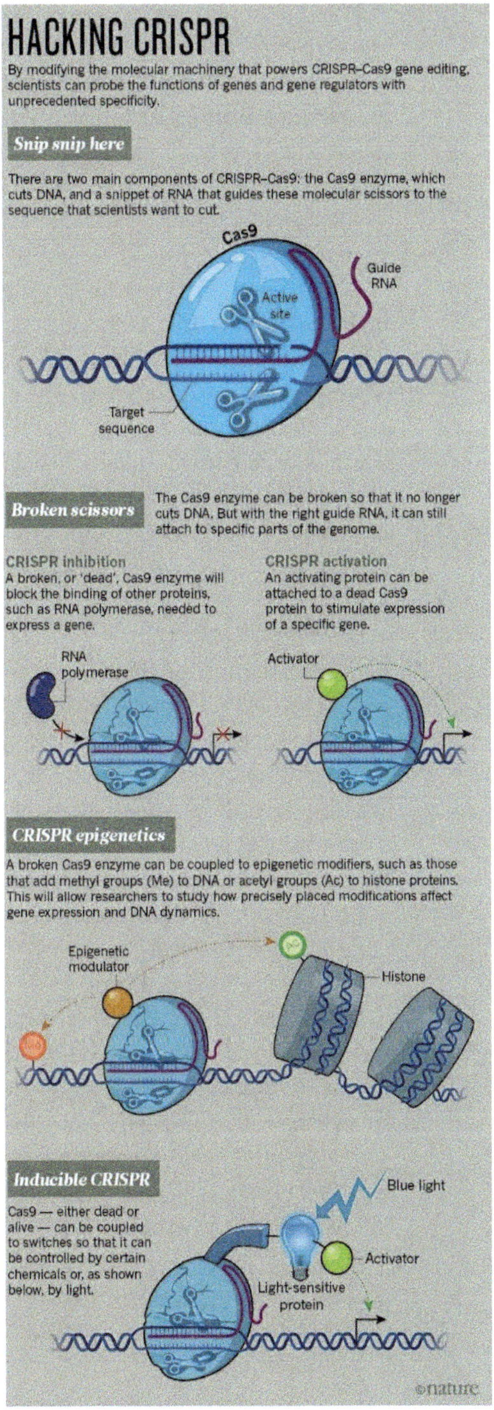

In addition, the article features six thumbnail pictures (three photos, a tiny world map, and two colorful drawings of DNA strands) that are actually links to other CRISPR-focused articles on the *Nature* news webpage.

Briefly using the rhetorical analytical categories introduced in Chapter One, we can suggest how these visuals work in the context of this article.

PURPOSES

The cartoon attracts a wide range of potential readers by the strange image of scientists on surfboards, but the overall serious **tone** of the article shows that the cartoon is only a lure. The article does at times reflect the light-hearted tone of the cartoon in its use of enthusiastic comments by participants in the research ("It's so fun!"), but mostly it attempts as clearly as possible to describe the complex technical processes of CRISPR for a wide audience by using the metaphorical language of "scissors," "cutting," "editing," and "tethering."

The infographic, in contrast, is a visual summary of the sections of the entire article. While its bright colors and its icons of scissors and DNA strands are meant to attract a varied readership, it summarizes complex processes that would otherwise be understandable only to scientific specialist readers. The infographic therefore does not substitute for reading the article, but may help readers decide which sections of the whole they may be most interested in reading. In this way, the infographic works almost as an abstract, although it does not make the CRISPR processes easier to understand.

INTENDED READERS

The news section of the *Nature* website characteristically uses colorful graphics to attract a wide audience of readers with varying levels of familiarity with a topic. It also uses written text that features widely familiar metaphors to lessen to some extent the technical difficulty of processes that are known well only by scientists in the specific research field. In this instance, most adult readers can understand the notions of "cutting" and "editing," but how these actions work in a lab setting, with enzymes as "scissors" and RNA as "guide" at a microcellular level, will remain unknown and largely unfathomable to non-specialists.

On the other hand, a wide range of readers—from scientists in other specialties to potential investors in research and to users of health services—will appreciate from such articles the **evidence** of the multiple studies cited and the corroborating testimony of other researchers and stakeholders. Note, for example, that the article begins not with a description of CRISPR, but with a description of an enthusiastic company "where study authors often deposit molecular tools that they used in their work, and where other scientists immediately turn to get them."

Exercise 3C: Select a recent *online* STEM article in an area of research of interest to you. List the drawings and diagrams in that article and do a rhetorical analysis similar to that shown just above. Be sure to analyze purposes and intended readers and be aware of how evidence and tone are used in the article to affect readers. Note specific terms, including metaphors (like "cutting" and "tethering" in relation to CRISPR) that are used to reach readers.

Do you feel that these metaphors or analogies help non-specialist readers understand the processes being described?

Focusing on a specific drawing or diagram, analyze its relationship to the text of the article. Does it help you understand the text? Does it augment information in the text? Does it perhaps confuse what was clear in the text? Does it reinforce the tone of the article or does it create contrast?

Lastly, how would you rate the rhetorical effectiveness of the drawing or diagram? How would you change it (or replace it) to increase effectiveness?

VIII. VIDEO AND AUDIO

Movies and video have increasingly pervaded STEM communication since science lessons were first televised in the early 1950s. As with photography, video capability has become personal and ubiquitous, with hand-held devices becoming increasingly powerful. We are also seeing advances in AI-generated video content that will shape how science is communicated and merits much consideration for its accuracy and ethics.

In the everyday *doing* of science, video recording cuts across all fields as researchers observe processes of all kinds. In fields such as animal science and ecology, both video and audio technologies are used to observe movement and processes that involve sight and sound.

As with the technology of linking to sources, video and audio use in STEM communication is restricted to online and other digital environments. Because print publication of peer-reviewed research does not support these media, and because print publication is still considered more prestigious in some academic circles—despite its limitations—peer-reviewed journals have been much slower to incorporate video and audio than have online and other digital STEM media. However, some journals now include video abstracts in the online versions of articles.

The most innovative work in STEM communication occurs on websites, and it is becoming more and more common for a print article to refer readers to the publication's website in order to get a fuller and more active sense of the research. This need for movement by readers from a print source to an online counterpart has implications for STEM rhetoric. In effect, the writer in the print environment is forced by the limitations of the print platform to either

- use words and numbers only (or perhaps with photography or drawings) to attempt to describe phenomena that are better shown in video or audio, or
- ask the reader to move to the online source rather than stay on the print version.

This clumsy shifting risks losing the reader's attention and interest. More importantly, a writer's striving for print publication can act as a disincentive to trying to describe video or audio data that might be more persuasive with intended readers.

A good example of how STEM communicators try to handle the emergence of video into the dominant print paradigm is the following article from *PLOS One* about the controversial "mirror and mark tests" performed on birds and primates to see if they show self-recognition, which formerly had been thought to be a trait only of humans. Observe how Soler et al. (2014) describe the physical movements of magpies in front of mirrors and non-reflective panels in tests conducted by Prior et al. (2008):

> In a carefully designed and well-controlled experiment, [they] found that magpies confronted with their mirror-image at the beginning responded with social behaviour (aggressive and submissive displays) and exploration of the mirror (approaching it and looking behind it), but later showed self-contingent behaviour (i.e., rapid left and right or back and forth movements in front of the mirror). After 250 min of cumulative exposure to the mirror, each magpie was subjected to eight sessions of the mark test, twice on each of four different conditions (as described in Prior et al. [2008]). In four of the sessions they were provided with a brightly coloured sticker and in the other four with a black (sham) sticker. Similarly, four of the sessions were performed with a mirror in the cage, and the other four with a non-reflective plate of the same size in the cage. Two out of five magpies were reported to pass the mark test. The sticker was stuck under the beak, in the throat region, outside the magpies' visual field, but these two magpies were capable of removing the sticker by scratching with their foot in mirror-present sessions. When the magpies were tested with the non-reflective plate, evidence of sticker-directed behaviour was negligible.

In a partial effort to accommodate the video technology that scientists since Prior et al. have used in mirror-test research, Soler et al. (2014) include, under "Supporting Information," links to PDFs of eight videos of bird tests.

Note that these links are not to the videos themselves, even though the *PLOS One* article is also online. The print paradigm is still dominant in this hybrid accommodation, but that is changing.

In contrast, excerpted footage (55 seconds) from videos by Prior et al. was published on the *New Scientist* website in a short summary article by Alison Motluk (2008) entitled "Mirror Test Shows Magpies Aren't So Bird-Brained."[4] Note how Motluk combines video with her audio interpretation of what the viewer is seeing. Here the video captures the viewer's attention. Meanwhile, Motluk's commentary, along with the editing of the video footage, attempts to direct the viewer's understanding of the video by putting it in the context of (1) what the researchers did to set up the test, (2) other research on magpies, and (3) their conclusions from the tests—all in 55 seconds.

Exercise 3D: Watch the Motluk video/audio rhetorically. Think particularly about the merging of video and audio. How does the combination work to achieve the writer's purposes? What audiences do you think Motluk is trying to reach? Does she do so effectively?

How might you change the video/audio to increase its effectiveness? What changes would you make? Why?

In further contrast, consider a somewhat longer video/audio blend (over four minutes) by J. Patrick Malone (2009), also about the mirror test, but this time recording chimpanzees.

The longest portion of the video excerpts recordings of the work of Jane Goodall in Tanzania. In this video, the audio introduces and contextualizes the longer video excerpts (as in the Motluk video), but the Malone video leaves a significant portion free of audio commentary, so the viewer concentrates on the video.

Exercise 3E: Consider how this different use of the tools and a longer timeframe influence rhetorical effectiveness. In particular, how would an audio-free portion influence your attention as a viewer? What is gained (and perhaps lost?) through these differences? Why?

Again, how might you use the video and audio tools to change the rhetorical situation and affect the value of the video/audio blend?

4 The excerpted footage can be viewed at https://www.newscientist.com/article/dn14552-mirror-test-shows-magpies-arent-so-bird-brained.

IX. STEM COMMUNICATION ONLINE

Platforms such as *PLOS One* may be "born digital," but many of their features are still strongly rooted in traditional scientific publishing. Other platforms offer a different approach to communicating science. Consider YouTube or TikTok, where videos can be shared and commented upon and where content is minimally supervised. Nowadays, almost anyone with access to a computer or cellphone can become a digital content creator.

Rhetorically speaking, we need to think differently about content creation in these spaces. Unlike traditional scientific publishing, it is difficult to know who the audience is for content. It is also difficult to know who the content creators are. This becomes a challenge when we think about sharing or reposting content. In these online spaces, we cannot be easily guided by

1. known **professional standards**,
2. the kind of explicit **guidelines** that peer-reviewed journals provide, or
3. the small range of **models** produced for writers in newspapers and magazines.

While there might not be standards, guidelines, or models in the same sense as more conventional forms of scientific publishing, one can use a wealth of rhetorical strategies.

When we think about communicating science, we might draw a continuum from formal genres used for discussion among scientists to popularizations. It is common to think about these two forms—internal and external audiences—as a binary, but there is much more between these two poles. When thinking about how science is communicated in online environments like TikTok or YouTube, we see a full range of genre uses. Researchers share highly technical videos for other researchers or students, some scientists work to popularize some area, and others use the platforms to share **urgent information**—as many public health researchers did during the COVID-19 pandemic.

There are uses of online tools in scientific research that draw on rhetorical strategies of these born digital platforms. Consider crowdfunding proposals used to help students and researchers undertake their work. Websites such as Experiment.com are like GoFundMe or Kickstarter, but for scientific research. There are also citizen science projects, like Foldit or the Great Backyard Bird Count, whose wealth of rhetorical resources help everyday people advance scientific research.

It might seem overwhelming to think about all these possible uses. But you can use your rhetorical insights about audience to shape how you tackle a new communication space. As a communicator, when crafting messages for these digital spaces, you will need to identify a readership whose expectations you do know and which are important to you, and address them.

Adopting this principle gets you beyond the bewilderment of trying to imagine the "general public" and provides a solid foundation for your knowledge of readers upon which you can base your design.

You can then begin to think about the norms for the platform that you are using. Communicating on TikTok, for instance, entails different audience expectations than doing so on YouTube, even. Understanding the space requires you to engage content other than your own and to look for exemplars. In these environments, genre change can happen quickly, and so you will need to familiarize yourself with current practices and keep up-to-date as long as you wish to use that platform.

THE RHETORICAL VALUE OF COMMENTS IN INTERACTIVE WEBSITES

Having noted the importance to the writer of addressing your work to a known readership, let's go back briefly to recognize the value of the open **comment function** on interactive platforms. As noted in Chapter Two when describing the need for feedback for relatively new writers, we emphasize again the opportunity that such comments help writers understand readers and develop a richer sense of potential audiences. If you have used social media to share content, you have likely also taken part in online conversations on such sites. In fact, you may have already become a rhetorical reader of the commentary on sites of interest to you. Even if the comments are anonymous, as they often are (or venomous, as they sometimes are), they tend to quickly give a reader a good sense of the range of perspectives and opinions and even backgrounds of the respondents—as well as the level of language that respondents use, their favorite metaphors, and what they respect as evidence. This is highly valuable feedback for potential writers as they seek to understand readers and the genres in which one is participating.

One should proceed with caution, however, when looking at online comments. First, we know from research and the practical experience of reading comments online that they can include vituperative attacks—racist, sexist, and other forms of discrimination. Further, with increasing click farms, bots, AI tools, etc., it is difficult to assess if comments are sincere or even posted by humans. Being rhetorical means that one critically assesses what they are reading with such suasive actors and their various agendas in mind.

Exercise 3F: Select a STEM topic of interest to you—one for which you are either

- aware of video resources online (which you could cite), or
- interested in creating a short video to accompany a report on research.

Whichever option you choose, be sure to consider the **six rhetorical categories** in deciding which video resources to include or the characteristics of the video you create. How does the chosen or created video help you achieve your purposes communicating this research? How does the audio portion contribute to the effectiveness? What readers would benefit from this video and why? Consider presenting this video (chosen or created) to fellow students or researchers as part of a short presentation of your research interest.

With the video, include a brief (e.g., no more than 500 words) explanation of what you hope to accomplish rhetorically by using this video. Writing this explanation will help you think about your communication goals and how you are trying to reach them. If the video you have chosen or created does not meet all your expectations, be sure to include in your explanation how you think it still needs to be improved. (Keep in mind that if you are creating a video, and are not experienced in this medium, you're likely to fall short of your expectations. Just consider this good practice!)

CONCLUSION: YOU AND THE PRESENTATION TECHNOLOGIES

The underlying message of this chapter—and of the book as a whole—is that we can't go back in technological time in our communication of science any more than we can, or would want to, go back in how we do science. The multimodal and more accessible communication of STEM-relevant ideas, data, and opinions is here to stay and with a growing number and range of people who have a stake in this communicative flow. The rise of participatory science movements—such as citizen science, community science, and community monitoring projects—has been one way to pay homage to this huge growth of respect for how those interested in science, but often without specialist training, can genuinely contribute to STEM projects.

There are also challenges, however, that come with the seemingly unlimited and unrestricted social media technologies, as anyone might comment with opinions of all kinds. Some may be valid and offer important distinct perspectives, but we must also contend with poor quality or inaccurate information. In these cases, we can talk about the importance of recognizing scientific, technical, or health misinformation and think about how we can each do our part to ensure our contributions are accurate, ethical, and fair.

Learning, as many of us are, to contend with the rhetorical challenges of this new and changing communicative world is increasingly part of what it means to do science.

CHAPTER FOUR

WRITING SCIENCE ETHICALLY

Topics in this chapter:

I. Covering Up Incomplete or Poorly Done Research, or Conflicts of Interest
II. Plagiarism
III. Ethical Issues and Artificial Intelligence
IV. What Is Common Knowledge?
V. Claims and Overclaims: The Dangers of Hype
VI. Striving for Accuracy in Language

I. COVERING UP INCOMPLETE OR POORLY DONE RESEARCH, OR CONFLICTS OF INTEREST

When a written artifact is found to be **unethical**, particularly when it has been published by a well-known and respected journal or other venue, *ethos* is damaged, and *everyone involved*—not only the unethical researchers—loses the trust of readers. As explored in Chapters One and Three, establishing ethos (i.e., reputation or character) is fundamental to rhetorical effectiveness. Maintaining that positive reputation is equally fundamental and takes consistent work by researchers and publishers.

However, discredited research—especially if originally published in a prestigious journal—can also have the effect of creating a **group of readers who believe the discredited claims**.

The effects can be far reaching, indeed. One of the best-known examples of how large the consequences can be is the highly politicized 1998 article by A.J. Wakefield et al., originally published in *The Lancet*, titled

"Ileal-Lymphoid-Nodular Hyperplasia, Non-Specific Colitis, and Pervasive Developmental Disorder in Children" (reprinted, with commentary, in Nelson-McDermott et al. 2014). This article led to, as the editors put it, "widespread fears that the measles, mumps, and rubella (MMR) vaccine could cause autism spectrum disorders" (183).

Although the claims of the connection between MMR and autism have been disproved by many subsequent studies, and although the article was eventually retracted from *The Lancet* in 2010, the fears sparked by the erroneous research have continued, with some parents refusing to have their children vaccinated—despite outbreaks of vaccine-preventable diseases in communities where significant numbers of children have not been vaccinated. The misinformation has also spread despite emphatic research-based statements by the UK General Medical Council that there is no connection between the most common vaccines for young children and the incidence of autism spectrum disorder.[1]

So, even if writing is not the *cause* of the ethical breach, the publication of written work based on bad science can *spread and perpetuate* the effects of unethical or incompetent behavior.

The Lancet example falls into a category of unethical science writing that can be called "covering up incomplete or poorly done research." In the same category can go such ethical breaches as failing to mention a potential conflict of interest by a researcher (as was also the case in the anti-vaccine example described above). In many instances, writers have been well aware of how they have misrepresented (lied about) methods or results. Ethical science writers place accuracy and truthfulness above all other considerations when reporting what has occurred.

Where, however, determining accuracy gets complicated is in being sure that writers have been complete and thorough in reporting research. Ethical writers constantly worry that they might have left out important details that may later be interpreted as an ethical lapse. Anyone who writes research articles knows the difficulty of trying to be complete in reporting *all* that is meaningful or relevant in a study. The most conscientious researcher, regardless of field, will struggle with

- how much background to include in an introduction,
- how many predecessors and studies to cite,
- how many details of methodology to describe,
- how many and which results to state, and
- how many and which possible conclusions to draw.

1 For a history of the incident, see https://en.wikipedia.org/wiki/Lancet_MMR_autism_fraud (Aug. 6, 2025).

Although striving for completeness, the ethical science writer, in writing in proposals and articles, always faces such constraints as word limits. Moreover, as we saw in Chapter One, the rhetorically savvy writer knows that obtaining and keeping the attention of readers requires that the writer establish a **focus** for an article and a clear sense of **purpose**, then choose the **evidence, organization, style and tone, and graphic presentation** that will lead to that purpose being achieved for the groups of readers the writer wants to influence. All of these factors put pressure on the researcher to be selective in what can be included in a draft for possible publication. Selectivity always means that some material must be left out.

Who is left out is another important consideration in the practice and communication of science. Science is not merely the discovery of objective facts, and science communication is not merely the delivery of those objective facts to different audiences. Most professional scientists are very aware of this and especially those who are familiar with the history of their fields. Science changes because it is a process of understanding the world, but that understanding is mediated by the limits of our own experiences, positions, histories, etc. Because science is a human process, the history of science can point to many failures to include or appropriately include everyone, and redressing these omissions is important work. When we think about communicating science, we ought to be mindful of its ongoing processes of discovery, of success and failure, and where omission of some people or voices may mean less complete insights.

COMPLETENESS IN PEER-REVIEWED JOURNAL ARTICLES

Fortunately, the studies and research reviews written for peer-reviewed journals usually give researchers room to be inclusive, and the standard formats of the research article and review (see Chapters Five through Seven) lead readers to expect some expansiveness in each section. Nevertheless, rhetorical effectiveness still demands **selectivity**.

COMPLETENESS IN STEM JOURNALISM AND SHORTER FORMS

In contrast, those writing for less-specialized and usually broader audiences in newspapers, blogs, websites, and magazines usually face sharper pressures to be selective and concise, so the ethical writer must achieve sharper focus and point the reader to other sources that can be more expansive. We've already seen in Chapters One and Three how this can work, as even those who work in very concise forms, such as news articles and video, can point readers to the peer-reviewed literature that provides the depth that the short forms lack.

The close relationship between science journalism and the peer-reviewed journal literature that supports popular literature in science is crucial to the rhetorical reach and effectiveness of science writing in general.

In other words, science journalists can afford to keep their articles incomplete—leaving out many details of the methods and results of the research on which they report—because they can refer (or link) their readers to the much longer peer-reviewed articles that first published the research. But, in doing so, these journalists do take the risk that many of their readers will not be able to understand the language of the specialist peer-reviewed journals, or may even skip them altogether. Still, journalists feel that this risk is worth taking, in exchange for the benefit of reporting scientific discoveries and controversies to a much wider audience.

TOOLS TO HELP WRITERS WRITE ETHICALLY

While acknowledging the pressures to be focused and selective, the STEM writer can achieve an ethical stance by

1. maintaining the **goal of completeness**, and
2. using the **writing process, particularly drafting and feedback**, to help choose what and how much to include in a draft to be submitted.

Maintaining the goal of completeness means that the writer

- continually strives for accuracy in describing research, its methods and materials, and the relevant results that support reasonable conclusions;
- will never deliberately cover over, omit, or misstate methodological flaws or omissions that could undermine the validity of results;
- will not use a word limit (e.g., in a proposal) or a time limit (e.g., in a presentation) as an excuse for misrepresenting the research; and
- will be sure, even when there are word or time limits, to note limitations in the research in the space or time available.

For example, being sure to mention briefly the size and characteristics of a study population or the number and types of tests run can keep readers from imagining that a study is larger or more comprehensive than it actually is.

Using the Writing Process to Aid the Writer in Being Properly Selective

When in doubt about what to include in a report or article and what to leave out, rely on the writing process—especially multiple drafting and opportunities for feedback—to help you make those hard choices.

If, for example, you have a word limitation for an article or a grant application, allow yourself to write a draft (or more than one) that greatly exceeds the word limit, and that includes everything that you might conceivably want to put into the document about your research. Use this draft as a tool to put into words (and appropriate images, tables, calculations, etc.) anything and everything that might turn out to be relevant in the report.

You'll find that you can write such a draft much more quickly and with less pressure than if you tried to write just one draft that tries to fit everything within the word limit. Mistakenly trying that "one-draft final" will just frustrate you and lead you to leave out much that might be more important than what you have included.

Once you have written the much longer and more inclusive draft, then you'll be in a good position to read through it carefully and begin to weed out what is less important. Continue with this process until you have a draft that both fits the word limit and satisfies you that you have included the most important and most accurate material.

Why is feedback important? Let's say that you have written that much longer first draft and are now not sure what should go and what should stay (or even be expanded!). Here is where another reader or readers besides yourself can help you make those tough choices (see Chapter Two, p. 60, for basic advice on getting feedback).

Who should give you feedback? These good readers should be people

1. who have some understanding of what you are trying to achieve, and
2. who are in some way representative of the audience you are trying to reach.

A fellow research team member could be a good choice. Another good choice would be a fellow writer who is experienced at giving feedback. The **goals of accuracy and completeness** are so important that you should not hesitate to ask for feedback from qualified readers, despite worrying that you may be impinging on their time.

How should I instruct or guide my reader? In asking for feedback, be sure to make it clear to your reader what purposes you are trying to achieve, the readers you are trying to reach, and why you need that reader's feedback. If you are trying to fit a 5,000-word draft into a 2,000-word limit, let the reader know that. If you have a very clear idea of the most important ideas you want to get across, let the reader know that, too. If you are worried that a specific description, table, or conclusion (or any other aspect of the draft) may mislead an audience, focus on that for feedback. Ask your reader questions about the parts of the draft that will help you resolve your quandary. Let your reader know that you are particularly interested in ethical considerations, and that you would like your reader, especially, to comment on any aspect of the draft that seems ambiguous or vague. Ask your reader to raise any concerns about accuracy. Listen closely to what your reader says or asks you. In this way, you can decide where in the draft you may need a better explanation.

Exercise 4A: If you are currently working on a piece of STEM-related writing, follow the advice given above about drafting for completeness. Then follow the advice about choosing a qualified reader to give you feedback, then the advice about guiding the reader.

Exercise 4B: Read through a piece of your STEM writing that you have previously written. Identify at least one part of the writing that you feel least confident about in terms of its accuracy or completeness. Write to yourself about the nature of your doubts. What do you think could be improved?

Then imagine yourself in conversation with a trusted reader whom you have asked to give you feedback. What would you tell your reader about your concerns regarding this portion of your writing? Now imagine that the reader has read your writing. What questions might you ask this reader about this portion of the draft to help you clarify or expand the writing?

II. PLAGIARISM

Plagiarism means using in your own writing the written words, processes, or ideas (or images, tables, etc.) of another researcher **without explicit attribution** or, in some cases, **explicit permission** from a publisher. If you wish to use more than a few paragraphs or pages from another copyrighted source in your work, you may need to acquire written permission from the publisher.

You can also plagiarize your own work! If you, for example, simply took a paper you wrote for one course and used it in another, this would be self-plagiarism if not disclosed and approved of by your instructor.

Plagiarism does not only mean word-for-word or cut-and-paste copying, but substantively borrowing words, ideas, etc. For instance, if you were asked to summarize a research article, you would need to rewrite it completely with your own words (paraphrasing), with its key ideas (summarizing), and then provide a citation. If there are highly specific words or phrases that you need to replicate in your summary, you would use quotations marks for those and then cite them (including specific reference to page numbers). Rearranging words, using similar words as a replacement, and so on is not sufficient to avoid plagiarism.

Especially in peer-reviewed journal articles, you must cite sources for anything you write that is not your own current research. You need not cite sources for what might be considered common knowledge. (See "What Is Common Knowledge?" later in the chapter. For citation practices in articles reporting new research in peer-reviewed STEM journals, see Chapter Five, and for citation practices in research reviews in peer-reviewed STEM journals, see Chapter Seven.)

More recently, artificial intelligence (AI) technologies have introduced complications in terms of how we think about plagiarism. These technologies are being used to summarize, to generate papers, and so on. But, as we will later discuss, their ethical use requires careful reflection. In most cases, your instructor will have an AI policy. Scientific journals are also developing policies and grant agencies and other bodies are also contending with what AI technologies mean for science. You must carefully consider these technologies and your use of them within the context of these fast-evolving conversations, expectations, and norms.

SELF-PLAGIARISM

Even if you include your own earlier published research in your research article, you must cite those publications in order not to mislead readers into thinking that all of the work in the article is new research. Not citing your earlier, relevant research is called "self-plagiarism" and can be a serious ethical breach, especially if it causes you to

1. **fail to credit** other publications,
2. **misrepresent** the currency of your research, or
3. **ignore the work of fellow researchers** who contributed to the earlier articles.

FAILING TO CITE SOURCES OF DATA

Moreover, with the increasing availability of accessible datasets online, avoiding plagiarism also means carefully citing the sources of any data you might use that was collected by other researchers.

> ***The Bottom Line: Accurately and thoroughly citing all sources used in your research and in the writing of that research is your best safeguard against an accusation of plagiarism.***

PARAPHRASING

Even if you are scrupulous in citing all your sources, you must also be careful to avoid copying into your own writing the exact language of the texts of those sources, even if you have cited them. If you are writing about your research for a blog, newspaper, or popular magazine, you can easily avoid this problem if you use **quotation marks** or **indented text boxes** to show that the quoted matter comes from a different source.

However, the solution is not so easy for those writing for peer-reviewed STEM journals. Why? Because it has become stylistic tradition in peer-reviewed STEM journals not to quote verbatim (with quotation marks) from other sources. This stylistic tradition puts STEM writers in a bind when they want to cite material from other articles. The writers can try to **paraphrase** the specific content from the other source, which means **using different wording than in the original while maintaining the same meaning as the original.**[2]

However, any paraphrase—because it changes the wording of the original—runs the risk of inaccurately representing the cited source. When writing for peer-reviewed journals, writers should think carefully about the need to paraphrase the sources they cite. In some cases, a paraphrase may be necessary, as in a methods section in which the writer is describing step by step a key process. If previous research is being cited for these methods, it may be necessary to cite the steps of that prior research, and so paraphrase can be used. However, in other instances, it may be sufficient to point the reader to the prior research and specific pages where the methods in question are described. In such cases, a short summary of the prior research (see the next section, on summarizing) will suffice to identify the prior research and point the reader toward it.

Keep in mind that in most cases an inaccurate paraphrase is less a matter of plagiarism—since the source is being explicitly credited—than it is of accurate representation of the source materials.

The risk of inaccurate representation is particularly acute if one wants to paraphrase highly technical material that one does not fully understand. Roig (2015, 11–12), for example, recommends the following:

> The ability to properly paraphrase technical text depends in large part on an author's conceptual understanding of the material and his/her mastery and command of the language and of her knowledge of, and ability, to convey discipline-specific expressions typically used to describe relevant phenomena, laboratory processes and procedures, etc.

2 See Roig (2015, 8–14) for a lengthy analysis of "appropriate" and "inappropriate" paraphrases, demonstrating the difficulty of achieving effective paraphrases of STEM articles.

You should never view either paraphrase or direct quotation as a substitute for your full understanding of what another person has written. Too many inexperienced writers copy into their prose what seems to be the authoritative language of others for the very reason that they don't really understand what the other person has written. Once again, if you are in doubt about the accuracy of your understanding of the sources that you are using in your research and writing, ask for feedback from a qualified reader regarding your paraphrases (or summaries) of others' writings.

SUMMARIZING

Whether in writing for peer-reviewed journals or in writing in other genres, you will frequently need to summarize the content of other sources, which means using far less wording to give the essence of the original content.

Typically, newspaper and popular magazine accounts must rely on summary for these reasons:

- The brevity of these articles necessitates great compression of the source text and severe selectivity of the content to be reported.
- The writer of the popular article usually assumes the intended readers will be interested in the consequences of the research rather than the details of methods and data, so the writer sharply focuses the summary of the source text on the effects that the intended readers will be most interested in.

From the standpoint of ethics, these factors mean that the popular science writer should be careful to represent the content of the source accurately and to avoid going beyond the researchers themselves in what they claim about the impact of the research. (See "Claims and Overclaims" below.) Note that in assessing the impact of the rescue research in Australia described in Chapter One, the *New York Times* authors support the projections of the benefits of the new methods with quotations from interviews with other authorities.

In addition, from the perspective of avoiding plagiarism, the authors cover the basics of ethical use of their sources:

- They **credit the several sources** (through the use of the links embedded in the article and by mentioning the research teams) so that interested readers can follow up if they wish.
- They **summarize** the content of the source studies **without directly copying** sentences from the source texts; had they wished to use some sentences, they would have placed them in quotation marks or in indented text.

Exercise 4C: Imagine that you are writing a blog entry or short newspaper article about a published research study in your area of interest. After you have read the source study and have it before you, try writing a very short summary of it (<200 words). Before you write, consider these factors:

- What background, results, or conclusions from the study do you want to emphasize?
- Who are your intended readers?
- What would you like these readers to learn from your summary?
- In order to represent the source accurately, what facts from the study do you want to be sure to include?
- In order to represent the source accurately, are there any misleading impressions about the source that you want to be sure *not* to communicate to your reader?

If you are doing this exercise in a group setting, compare your draft of the summary with at least one other writer's draft. What is different and similar about the two summaries? How do you account for the differences? What can you learn from the other writer's summary?

III. ETHICAL ISSUES AND ARTIFICIAL INTELLIGENCE

Although AI technologies have been in use for many years, their ubiquity has been recently marked by the release of OpenAI's ChatGPT and similar tools. These AI tools allow for content to be generated through predicting what would be a common word, for instance, to follow another. In many ways, these technologies are incredible feats of engineering, but we must keep in mind that they are not, in fact, creating knowledge. We will return to this point in a moment, but we must also note that the rapid pace of development has seen enormous gains since the release of ChatGPT. OpenAI, which has already demonstrated a tool that turns text inputs into impressive videos, is only one of many companies developing tools, and we see AI systems now being integrated into a wide range of applications, including assistive writing apps.

The ethics of AI are far too complex to treat in full here. Before using these tools, one should consider how one feels about how these systems are built and trained: they sometimes use copyrighted content; the web is scraped for training data, which raises issues of content quality informing the systems and copyright; and there are also concerns about human beings tasked with data cleaning, including removing profoundly disturbing content, and the impacts on people of that exposure to harmful content. Because the data used to train systems are also already shaped by existing values, some existing vices are also baked into the systems, including sexist or racist beliefs. We must also consider how our own data are being used to create these systems and what rights we have to manage that use.

For example, consider the development of facial recognition technologies, how your likeness may be used, and what applications these technologies may have for surveillance. These serious ethical considerations should inform your decision to use AI and, if you choose to, how you vet what technology or application you are comfortable using.

Issues most directly related to writing and communicating for STEM are related to the use of AI for scientific research, for sharing science with broader audiences, and also related to challenges posed by misinformation.

Researchers have used AI to write at least parts of their articles (Kobak et al. 2025). We know this because some papers have included AI prompts repeated back in the output. Scientists and editors of scientific journals are very concerned about this use because AI does not *know* anything. The predictive way that text is generated is not the same as having knowledge. For example, sometimes AI just makes up a citation!

Although we are still in the early days of AI technologies, and while these are making a significant impact, many scientists are worried. When we think about ethics in our communication practices as scientists, we think about what we write, what we review, and what we share. Writing with AI tools asks us to consider our practices in writing and composition, including

- how we generate ideas,
- how we think about and describe our audience,
- how we think about our own writing process,
- how we revise and improve our arguments, and
- what unique voice or style we have.

These practices are shaped by what we know, the experiences we have had, and those individual features that allow us to bring information together in a coherent form we call knowledge. These are rhetorical and ethical practices. We must also consider our duty to our work, to our colleagues, and to our audiences, and must thus reflect on the appropriateness of using AI tools in writing our own work or reviewing the work of others.

Disinformation and misinformation have long been discussed, but advances in AI supercharge these forms of disordered information.

Disinformation means inaccurate or completely wrong information that is **intentionally fabricated**, and **misinformation** means inaccurate or wrong information that is **not necessarily intentional**. For example, someone might accidently share inaccurate messaging about so-called climate cycles, which they believe to be true, but which are not scientifically accurate.

Some people, however, set out to create and share false information. Disinformation is a real problem because it creates confusion, and tracking down sources and vetting all the information that we encounter in a single day would be exhausting—even if we could find enough time to do it. With advances in AI, creating disinformation is much easier, as these systems can quickly produce text, images, and even video that does not reflect scientific evidence or even reality. The volume of content that can be created is a serious concern, and efforts to combat disinformation and misinformation in science have required significant human and financial resources. Science communicators must be aware of common disinformation and misinformation in their area of work so that they can enter conversations with an awareness of common questions, concerns, or confusions.

Addressing incorrect information is not merely an exercise in sharing more facts, data, or "correcting" someone. In science communication, we often talk about the **deficit model** or the idea that someone simply has a deficit in information or knowledge, and that if we simply fill that gap they will change their minds. As rhetorically aware communicators, we enter conversations to build bridges and understand the concerns someone may raise.

Let's consider a common and contentious subject: vaccination. Throughout COVID-19, "anti-vax" positions were often criticized, but it is important to understand the motivations of anti-vaxxers and to provide an honest effort to earn trust. If a parent is concerned about vaccination for their child, simply rattling off facts is unlikely to be persuasive; on the other hand, detailing the ways in which vaccines can help protect their child while providing a realistic account of adverse events and also talking about their specific concerns is being more rhetorically aware. Building ethos and trust, however, can take time, and sometimes we are not the best person to communicate with every audience. A rhetorically sophisticated science communicator will also think about where community and trust already exist, such as with family practitioners, pharmacists, and even community leaders who can help address people's concerns about vaccines.

(**Note**: Due to space limitations, we cannot provide a full account of plagiarism nor a thorough explanation of all the elements of possible plagiarism; hence we recommend you become familiar with more thorough guides to plagiarism in STEM writing, such as Roig [2015]. Regarding AI, because of the rapid speed of technological development, it is difficult to anticipate any and all changes; hence you should consult journals, your institution, or journalistic guidelines to keep abreast of current best practices.)

IV. WHAT IS COMMON KNOWLEDGE?

Simply put, **common knowledge** in STEM contexts is information that you can safely assume that **your intended readers know**—and for which they **would not expect you to cite** a source or sources. But *caution*: what is common knowledge to one group of readers will not be common to a different readership. For example, while you would not need to cite a source for a description of subatomic particles to an audience of physicists, you might need a source for that description if you are writing about new work at CERN, the European Organization for Nuclear Research, for an audience of science generalists. One way to think about whether something is common knowledge or not is to ask yourself, "Is this idea or detail so well known that I could find it without citation in many works written for this audience?" If you can easily answer "yes" to that question, then most likely you won't need to cite a source.

Conversely, if you know that an idea, concept, or opinion is controversial or new—and especially if you can associate that idea with some specific researchers—then you must cite a source or sources.

If you are in doubt as to whether an idea or detail is common knowledge, **the safe course is to cite**. You can always rely on your qualified draft reader to give you advice on whether you have provided citations for common knowledge, or you can rely on the editor of the publication that accepts your work to mark when an item for which you have cited a source can be considered common knowledge.

V. CLAIMS AND OVERCLAIMS: THE DANGERS OF HYPE

Amid constant social media image-building and aggressive advertising everywhere in our lives, the greatest ethical threat to scientific thinking is overclaiming—hype, hype, and more hype!

The classic image of the scientist—alone in a dreary lab, silently and meticulously poring over inscrutable notes, peering into a microscope, or running, for the umpteenth time, the same obscure test to be sure of consistent results—seems a laughable anachronism. Fundamental science is often sidelined and applied research is not only overemphasized but often overstated in its current application in media reporting. For example, we have discussed the emerging AI technologies and, while they do hold promise, hype around their application is everywhere.

Meanwhile, some scientific achievements, like the painstaking experiments in multiple disciplines that have contributed bit by bit to the overwhelming evidence for human-caused climate change, feel like a dismal downer to the millions who just want science to be about the feel-good, the new, the ever glitzier. Already, many are making claims about how AI technologies will help us address climate change, but we already know what steps need to be taken and it is now a policy, not a scientific question, that is needed to make climate action happen. We know those steps because of the careful, patient research scientists have conducted for decades.

We might also consider the example of mRNA vaccine developments in response to the COVID-19 pandemic. There has been much discussion and debate about vaccines, and there was indeed hype. While their rapid development was an astounding testament to the state of science and global scientific collaborations, the hype around the seemingly rapid development of the mRNA vaccine backfired in some quarters. mRNA vaccine technologies are not new: they were in-development and even tested as early as the 1990s in animals. The Johns Hopkins Bloomberg School of Public Health does an effective job of detailing the development of these technologies in a webpage[3] where you will find a brief Q&A and embedded videos in the form of an explainer, which begins with a question that alludes to but does not directly reference some of the hype:

> Messenger RNA, or mRNA, was discovered in the early 1960s; research into how mRNA could be delivered into cells was developed in the 1970s. So, why did it take until the global COVID-19 pandemic of 2020 for the first mRNA vaccine to be brought to market? (Beyrer 2021)

The explainer tells us the history and does a good job of avoiding hype while relating that bringing mRNA vaccines to market indicates an important step forward.

For scientists committed to open-minded, painstaking research that does not promise instant gratification, rhetoric comes in for its share of blame for the current frenzy. An age-old complaint about rhetoric, going back to the ancient Greeks, is, as the philosopher Plato (1871) wrote, that rhetoric is merely "the habit of a bold and ready wit, which knows how to manage mankind," but is not committed to truth or justice.

3 You can visit the website at https://publichealth.jhu.edu/2021/the-long-history-of-mrna-vaccines.

By thinking rhetorically, as we have been arguing from the outset, you think about your messages in terms of (1) the purposes you want to achieve, (2) the readers—audiences—you most want to influence, and (3) the evidence, language, and images you want to use to achieve those purposes for those readers. From this perspective, rhetoric can look pretty much like clever manipulation. It may seem as if whatever your purpose might be, no matter how self-serving and dangerous, rhetoric will help you get there. That's what Plato was complaining about.

That's why this chapter, "Writing Science Ethically," is so important to understanding STEM rhetoric. The rhetorical principle emphasized in this chapter is **ethos**, the character of the writer, which embodies ethics. To write science ethically is to live the ideal of **scientific practice**, that careful, often tedious, and open-minded process that **weighs evidence and does not jump to easy or the most attractive conclusions**. Whereas the public—and, too often, those who can fund scientific research—often clamor for quick results that "prove" an attractive hypothesis, like a so-called cure for a disease, ironically, it is only the ethos of the careful, patient, unbiased researcher that gives science the power to maintain the respect of people.

When scientists give in to the pressure to produce quick results and to hype their work as blatantly as any other advertiser would, that confidence and trust in science erodes. In an atmosphere of claims and counterclaims by purveyors who are equally loud and manipulative, readers and viewers are tossed back and forth from opinion to opinion, and the advantage that scientists formerly held because of their ethos disappears.

This loss of the scientist's ethical stature is especially challenged by online platforms that encourage click-bait style posting to drive up content views. While it is necessary to catch a reader's attention, one ought to consider the possibility of miscommunicating information and the downstream consequences of doing so.

So, how can we use the tools of twenty-first-century communication in a way that upholds the scientific ethos and that still reaches those billions of viewers?

Here is a list of principles and techniques that science writers can follow:

1. In STEM communication, always keep the link with **peer-reviewed, open-minded research**, the process that supports it, and the ethos of science visible.
2. On websites, in press releases, and in other genres meant for non-specialist readers, always mention the **limitations of research**, the limitations of its products, and the need for a **conscientious assessment process**.

3. In abstracts of peer-reviewed articles, and in the articles themselves, maintain a **cautious tone** that avoids an enthusiasm that might appear to be advertising.
4. In writing to different audiences about research—including audiences as different as fellow researchers and potential consumers of a product or procedure—maintain a cautious **consistency of tone and message** that does not raise unrealistic expectations.
5. In opinion writing (e.g., in newspapers or on broadcast media), to counter arguments that appear to you to be antagonistic of science, **avoid emotional or sarcastic rhetoric** that imitates the tone and style of the antagonists.

Let's explore each of these principles in more detail:

1. In STEM journalism, always keep the link with peer-reviewed, open-minded research, the process that supports it, and the ethos of science visible. Always refer to and make prominent the published peer-reviewed research on which a newspaper or magazine article is based. Moreover, note such details as

- the length of a research process,
- the size and characteristics of study populations,
- the approval process that might still remain, and
- the comments from other qualified researchers, including any that might ask for restraint in enthusiasm about the hopeful findings.

2. On websites, in press releases, and in other genres meant for non-specialist readers, always mention the limitations of research, the limitations of its products, and the need for a conscientious assessment process.

Consumers have become used to print ads for pharmaceuticals that feature vivid photos of smiling, happy people who have presumably taken the medication being advertised. These enthusiastic visuals are bolstered by enthusiastic texts proclaiming the wondrous effects of the drug. But then the ads also include—in very fine print—lengthy, legalistic descriptions in medical language of the side effects and potential dangers of the drug, as well as lists of individuals who should not take them. Since the US Food and Drug Administration in 1999 mandated to companies the dual reporting of benefits and side effects in advertising, companies have used this method of reporting in print advertising (see Russell 2015).

Meanwhile, in TV advertising, dual reporting usually includes musical videos of the smiling, happy people engaged in many fun-filled activities—while a monotone narrator very rapidly lists the side effects and other contraindications. The overall effect is of a radically split consciousness—happy/ sad—rather than an **integrated presentation** that expresses what we might call "cautious optimism." The happy/sad split forces readers and viewers to choose where to direct their attention—with the purely happy message being much more vivid and alluring. Is this method of reporting research ethical?

The cautious optimism approach would blend the positive research story with

- a realistic sense of the **obstacles** facing researchers,
- the **ambiguity** of results, and
- the ongoing need for **skepticism** in the verification process.

3. In abstracts of peer-reviewed articles, and in the articles themselves, maintain a cautious tone that avoids an enthusiasm that might appear to be advertising. An abstract should provide an overview of the research and, in some cases, journals will have a specific format to follow, or there may be disciplinary norms you can look to for guidance. For instance, in many sciences, you will roughly use the outline of the paper to structure your abstract, including the main question or argument, the methods used, the results of analysis, and perhaps a sentence about the conclusion your findings support.

4. In writing to different audiences about research—including audiences as different as fellow researchers and potential consumers of a product or procedure—maintain a cautious consistency of tone and message that does not raise unrealistic expectations.

While most of us have come to expect corporate advertising of the products of science (e.g., automobiles, skin-care products, pet foods, solar panels, etc.) to be unmitigated shouts for joy, we still do not expect research-team websites or government regulatory websites to have a "hard-sell" ethos.

Nevertheless, writing rhetorically may lead some STEM writers to imagine that they can craft one message to a researcher audience in peer-reviewed journals—a message that is fully aware of the need to note limitations of results and inadequacies of methods—but craft a very different message for potential consumers and investors—a message that emphasizes benefits and minimizes or ignores limitations.

Ethically, STEM writers are on very shaky ground if their prose and images raise unrealistic expectations for one group of readers, even if they are more cautious in their approach to STEM colleagues in peer-reviewed journals. STEM writers should assume

- that at least some readers of a non-specialist website will read the peer-reviewed literature and so be aware of inconsistent messages. The ever-increasing accessibility of online journals will only continue to increase the readership of peer-reviewed research.
- that readers of enthusiastically positive messaging will develop unrealistic expectations of research and its results. At a time when people have come to expect instant miracles from science and technology, STEM websites should ethically be sure to counter such expectations through
    - realistic assertion of likely timelines for development and approval,
    - realistic estimates of actual impacts (not projected, hoped-for outcomes at some future point), and
    - accurate descriptions of impediments to successful research.

5. In opinion writing (e.g., in newspapers or on broadcast media), to counter arguments that appear to you to be antagonistic of science, avoid emotional or sarcastic rhetoric that imitates the tone and style of the antagonists.

STEM communicators, whether in writing or in speech, such as in interviews and talks, should avoid responding stridently and simplistically to attacks from those they perceive to be antagonistic to science. While issues such as global warming and human-caused climate change are emotional for millions of people and therefore elicit strong opinions, a scientific perspective earns its respect and power to persuade from its commitment to careful, rational research and its devotion to accurate data.

Current social media platforms such as Twitter (known as "X" since 2023), with their drastic limitations on message size, encourage "sound bite" responses filled with venom and sarcasm. These may play well with those who share one's point of view, but they have little power to change the minds of those who believe otherwise, and they surely fuel further highly emotional attacks.

From an ethical standpoint, this emotional pseudo-writing undermines the ethos of science, which requires space and time (in the context of spoken and written communication) to develop analysis of data and to build convincing arguments for the ramifications of research.

Scientific organizations devoted to political action, such as the Union of

Concerned Scientists, have spent many years refining a rhetorical approach that (a) respects diversity of opinion while (b) using the results of research to bring about political change in favor of renewable energy and its technologies (among other issues). The Union's monthly newsletter, *Catalyst*, shows that it is willing to use sound-bite rhetoric on occasion—e.g., the slogan "Science. Evidence. Facts. Reason" (Shulman 2017)—but only when combined with longer genres, such as letters to legislators and agencies, journalistic narratives about local communities (for example, some affected by pollution and sea-level rise), or policy statements that summarize specific research projects.

Alternatively, become familiar with the non-specialist magazine literature that reports research—and that presents it in both written text that is accessible to a broad readership and through colorful photography and attractive infographics. Magazines such as *National Geographic* and *Scientific American* usually reject a strident, emotionalized style. They feature essays on topics of interest to many groups of readers worldwide. Through stories of research, researchers, and their adventures and workplaces, they reach a broad spectrum of readers—and enhance knowledge of and respect for the difficulty and persistence of scientific inquiry.

Chapters Eight through Eleven will explore the contemporary genres and publication sites that demonstrate ways that science can be reported accurately and responsibly to many groups of readers less likely to read the peer-reviewed literature.

Exercise 4D: Compairing Claims: Choose a pair of documents that focus on the same research project—for example, a peer-reviewed research report from a journal of your choice and a newspaper account of this research that cites the peer-reviewed article; OR a peer-reviewed research paper and the website of the lab that produced the paper.

Read each document of the pair you choose: as you read, compare the claims they make about the impact of the research. In the research report, pay particular attention to the abstract and to the conclusion (or the section marked 'discussion' or something similar). In the newspaper account or the website description of the research, pay attention to what is said about the potential impacts of the research.

Do the claims in each document differ? How? For example, does either document mention the limitations of the research methods or of the impacts of the research? Is the tone of one document more enthusiastic than the tone of the other? Do you feel that the way in which the research impacts are described might mislead either group of readers?

How could the writing in either or both documents be changed to achieve greater consistency in the messages—or are you satisfied with the consistency?

Exercise 4E: When Is Reporting of Results Advertising, Not Science?: Choose at least two newspaper or magazine accounts of research projects in a field of interest to you. The two accounts need not be about the same research, but each must refer to a peer-reviewed publication of the research described.

Compare the two (or more) accounts. How similar and different are they in terms of their enthusiasm in discussing the potential impacts of the research? Does either or do both appear to be advertising the results (or products) of the research to, say, potential funders of further research, investors in the products, or practitioners who might recommend them?

Does either or do both accounts mention any limitations of the research, or advise caution in interpreting the results, or note how much time may be needed to develop products from the research? How are these factors worded in each account?

With classmates or others who have also done this comparison, discuss the ethics of using research reports to "advertise" products, either indirectly or directly. What are arguments pro and con?

Exercise 4F: Expressing Political Opinions about Scientific Issues: Should We Fight Fire with Fire?: Science has always been controversial and politically charged, because it almost always offers new ideas that clash with accepted values—or differing interpretations of data that pit one group of scientists against another.

This chapter maintains that STEM writers and speakers have an ethical responsibility to argue in a way that upholds the commitment of science to careful gathering and analysis of evidence, and that recognizes the limitations of methods and the need to continually improve how we observe, analyze, and interpret. This view rejects responding to opponents through emotional or sarcastic sound bites (or tweets) that might feel good at the time, but that undermine the scientific process.

Discuss with others how scientists (such as yourself) might respond to political opponents or those who seem to oppose certain scientific ideas. Should scientists respond to these antagonists? If so, how? What kinds of responses seem ethical to you and your colleagues? What seems unethical?

Considering writing, especially. Think about the wide variety of ways in which STEM writers can make political statements—from tweets, to Facebook posts, to signs at marches, to cartoons, to petitions, to letters to Congress (and other bodies), to policy statements by groups—to the everyday work of carrying out publication projects in order to advance research. Are each and all of these viable? If so, how? Would you argue that any of them should not be used by scientists?

VI. STRIVING FOR ACCURACY IN LANGUAGE

We can probably all agree that scientists should *always* strive for accuracy and should *never* deliberately mislead readers with inaccuracies. Indeed, every section of this chapter thus far,

- avoiding cover-ups of bad research
- avoiding plagiarism
- identifying common knowledge
- avoiding overclaims and hype,

involves striving for accuracy in how we write and how we use the writing of others.

But *striving* for accuracy in language and *achieving* it consistently are very different things. Every chapter of this book asks you as a writer to develop a **rhetorical frame of mind**: one that is conscious of how **what we write will change** depending on the **purposes** for our writing and the **readers** we intend to reach. Writing that may be accurate and understandable by a readership of fellow scholars in our research specialty will likely be unintelligible to a less specialized readership.

We might say, "I'd rather be accurate than understandable" (and some scientists do say this), but then we are leaving it to others, who may not be as knowledgeable as we are, to translate our language into other language that will be intelligible to the readers who need to use our research. So, in reality, there is no escaping the need for STEM writers to be flexible and versatile in language, adapting as best we can to the needs of the various groups of readers who can benefit from our research.

And even if you as a researcher do choose to leave it to others—journalists, editors, translators, grant writers, publicists, marketers, graphic artists, etc.—to change your language into language that is useful to other audiences, you will still need to develop enough of an awareness of rhetorical possibilities to judge the value of how those others have manipulated your language. Whether you like it or not, you will need to become, to some degree, a scholar of language.

Thus far, we have presented some samples of actual language in scientific writing for you to read closely and analyze rhetorically. The exercises have pushed you to do structured reading and questioning of many other samples of your choice. Some exercises have asked you to adapt to your own writing what you have learned by analyzing rhetorically the pieces that you have read. The remaining chapters will continue this process of interplay between structured reading and your own application in writing.

Moreover, some will become more specific in terms of genres of STEM writing, while others will look specifically at elements of style and the process of editing. But all of them will deal with ways to help make your reading and writing—uses of language—more versatile and systematic. In other words, to help you develop that rhetorical sensibility and strive for an accuracy that is appropriate to whatever rhetorical situations you encounter.

In the remainder of this chapter on writing science ethically, keep in mind the relationship among

- striving to be ethical,
- accuracy, and
- rhetorical appropriateness.

To put this another way, ethical writing requires

- the intention to be ethical (striving),
- the achievement of some degree of accuracy, and
- the appropriateness of this accuracy in the rhetorical situation.

None of these requirements is easy, because each offers the writer a range of options:

- What **ethical action** is in a given situation may be difficult to determine—for example, as we discussed earlier, if a grant proposal requires that I write no more than x number of words, I must summarize my research, and this will mean that I must select which details to include and which to omit. This selectivity means that to some degree my explanation will not be complete and may in fact be interpreted as misleading. Hence, this requirement emphasizes not the achievement of ethical writing per se, but the striving to be ethical, which often entails making choices among options. The sections, examples, and exercises in this chapter have been set up to help writers make those difficult choices.

- **Achieving "some degree" of accuracy** creates a shifting standard and leaves room for much variety of performance. Nevertheless, it realistically reflects the difficulty *both* of measuring complex phenomena *and* of translating observations into acceptable language. For example, when we wrote about paraphrasing earlier in this chapter (p. 102), we noted that the most accurate rendering of the words of others would be to quote them verbatim; however, we also noted that traditional practice in peer-reviewed STEM journals does not allow direct quotations from sources. This tacit prohibition means that writers must paraphrase, summarize, or merely cite sources. Paraphrase and summary are inherently different from the original and hence not entirely accurate. Mere citation does not attempt accurate translation of an idea, but rather it directs readers toward the source itself. Still, if some degree of accuracy is our standard, then we can compare paraphrases and summaries to try to judge accuracy.
- **Appropriate language for the rhetorical situation** is the most ambiguous requirement of all—which is why books such as this one need to be written, and why they need to be updated as writing environments and technologies change. In Chapter Three, for example, we compared ways in which writers choose among different currently available media—e.g., words, mathematical language, charts/tables/graphs, photographs, video, drawings, etc.—to attempt to portray phenomena accurately and to allow intended groups of readers to understand and use the information. Similarly, in this chapter we have looked at samples of scientific language through an ethical lens and weighed them against the standard of science as a careful, patient, open-minded quest for truth. However, we have also questioned if that standard for ethics in science communication is still viable, or at least exclusive, when modern communications technologies such as Twitter, Snapchat, and Facebook favor rapid, brief, highly emotional messages. Is it possible to be an ethical STEM communicator when using these rapid-fire tools? Conversely, if we do not use them, are we cutting ourselves and huge groups of readers off from scientific knowledge and thought? And would that in itself be a betrayal of ethics?

Exercise 4G: Think about your own striving to be an ethical science communicator. Consider your history as a writer on STEM-related topics; include not only professional papers or, if you're a student, essays and labs done for classes, but also any other writing you've done, formal or informal, about your science interests. How have you tried to measure up to the three requirements of (1) striving to be ethical, (2) achieving some degree of accuracy, and (3) using language appropriate to the rhetorical situation?

What challenges have you faced? What pieces of writing stand out for these ethical challenges? How did you deal with those challenges? Did you ever fall short, or do less than your best, in writing ethically? Are there lessons you've learned that you try to apply in your writing now?

Use these questions to write about these issues for yourself. Then, if you're so inclined, write another draft that you'd feel comfortable sharing with other writers also writing on this topic (perhaps in a class setting or writing group). If your experience has taught you lessons about ethical writing that others might be interested in, consider sharing them.

Exercise 4H: Think of a STEM writer (or research team) whose writing you admire. Describe the characteristics of the writing that you find so positive.

Looking at this person's (or team's) written work through the lens of ethics, is your admiration based at least in part on how this work exemplifies ethical STEM writing?

Describe, in relation to a particular piece of the writing, how the three requirements (striving to be ethical, achieving some degree of accuracy, and using language appropriate to the rhetorical situation), may have been met in the piece of writing you've chosen.

Despite your admiration of this person's or team's writing, can you suggest how the piece of writing you've chosen to analyze could be either more accurate or rhetorically appropriate? Describe one or more changes you'd suggest, and why you think they should be made.

CONCLUSION

This chapter has explored a number of common forms of unethical STEM writing—and provided advice on avoiding these pitfalls. In the next chapter, we focus sharply on the peer-reviewed STEM research article and offer advice on ways to achieve rhetorical effectiveness in this important genre, which is the bedrock of advancement in science. Chapter Five is actually the first of three chapters devoted to this vital form:

- Chapter Five looks at introductions, methods and materials, and abstracts.
- Chapter Six considers results and discussions.
- Chapter Seven looks at the related genre of the research review.

In reading and doing the exercises in all three chapters, you should keep in mind the themes of this chapter on the ethics of science writing, because they apply to everything we consider from this point forward regarding rhetorical effectiveness.

CHAPTER FIVE

WRITING THE RESEARCH ARTICLE, PART ONE: ABSTRACT, INTRODUCTION, METHODS AND MATERIALS

Topics in this chapter:

I. Thinking Rhetorically about the Peer-Reviewed Research Article
II. Giving Momentum to Your Research "Story"
III. Writing the Abstract
IV. Writing the Introduction of the Full Article
V. Writing the Methods and Materials Section of the Full Article

I. THINKING RHETORICALLY ABOUT THE PEER-REVIEWED RESEARCH ARTICLE

The peer-reviewed journal article is the core genre in STEM publication: the bedrock, the foundation, the gold standard. All other STEM communication that we analyze in this book draws its power from the quality of the research—and the writing—that goes into the peer-reviewed literature. The IMRD structure of the peer-reviewed research article,

- **I**ntroduction
- **M**ethods and Materials
- **R**esults, and
- **D**iscussion

shapes how we think about the conduct and reporting of research. Each of the sections answers basic questions that intended readers will ask about any research project:

- **Introduction**. What is the specific area or topic of this research? Why do you do it? What relevant research came before? How does your research fit into that tradition? How does it differ from it? What are you trying to show? What questions are you trying to answer?
- **Methods and Materials**. How did you design the research? Why this way? Where did you conduct it? Why? What steps did you follow? Why? What tools/participants/tests did you employ? Why? How were results recorded and data secured? How were the data analyzed and their significance calculated? If there are human participants, how did you ensure their safety and privacy? Show me everything you did, so that I (or someone else) could replicate your research or adapt it. Were there steps you wanted to perform but couldn't? Why not?
- **Results**. What were your findings (or what have they been so far)? What have you selected to show us in this article and why? What forms do those results take: calculations, data points, graphed trends, transcripts, images, etc.? What data are you merely summarizing? Why? What types of data are being left out of this section? Why?
- **Discussion**. What do these findings mean to you? Do they answer your initial questions? Why and how? Or why not? What other kinds of significance might your results have? How confident are you about the significance you are claiming? How have you measured that confidence? What limitations or cautions should we acknowledge about what you have found? What are the next steps for this research? Why? What remains to be done?

You might consider using these questions as a template in writing your own articles. We'll return to these questions for the exercises later in this chapter (and in Chapter Six).

PURPOSES AND INTENDED READERS

Here we apply the basic rhetorical categories—**purposes** and **intended readers**—to the peer-reviewed article. Inexperienced writers have a tendency to think of the IMRD sections as just empty spaces to be filled in. But in fact they are highly purposeful tools that you can use to achieve important and specific purposes for your writing—and to influence the readers you identify as needing the information your research can provide.

Considering Purposes

The most basic purposes for your article are embodied in the questions listed above in the introduction section of the IMRD structure:

- Why are you doing this research?
- What are you trying to understand and demonstrate to others?
- What questions are you trying to answer?

If you can answer these three questions for yourself in a way that satisfies you, you have begun to build a strong sense of purposes for your article. These can be called **internal purposes**, since they derive from the **research itself** and from the **IMRD structure**.

But your answers are not sufficient in themselves to justify the writing of the article. You must augment your sense of internal purposes by thinking about the questions listed for the discussion section:

- What do these findings mean to you?
- What other kinds of significance might your findings have?
- What are the next steps for this research?

Being able to answer these questions in a way that feels honest and satisfying to you empowers your article well beyond your answers to the introduction questions. They give you strong reasons to share your research with others and to make it convincing to those readers.

Further Purposes—External

But even answering these questions will not be enough to explain what you hope to achieve by writing—and publishing—this article. Experienced STEM writers have multiple **external** reasons for wanting to publish their research; possible purposes include the following:

- To bring more attention and respect to themselves and to their subject area
- To encourage funders to invest in further research on this topic
- To bring attention and interest to products that might be generated by this research
- To bring attention to and create understanding of political and societal consequences of this research

Exercise 5A: If you are already an experienced and perhaps published STEM writer, jot a list for yourself of other purposes for your writing that are not on the list above. OR, if your main external reasons are already on this list, write specifically about how these reasons apply to your research and will influence how you write.

(In doing this exercise, you might focus your thoughts on a specific article you have recently written or are writing.)

External purposes for inexperienced STEM writers, including students. If you are *not* already an experienced STEM writer, or are perhaps a STEM student with research you will "write up" for a teacher or lab director ("PI," for principal investigator), your external purposes for writing within the IMRD structure may differ from those listed just above. You may, for example, be motivated primarily by the desire

- to earn a high grade from your instructor or PI,
- to earn a grad school or professional school recommendation,
- to land an internship or fellowship.

All of these are important motivations that are part and parcel of the scientific culture. If they are important to you, you need to be aware of them and consider how your writing will be influenced by them.

At the same time, you'll need to keep in mind the internal purposes that stem directly from the questions implied by the IMRD structure and by the specific questions listed above from the introduction and discussion sections. Regardless of the important external motivations you may have for succeeding with your STEM research writing, understanding the internal purposes implied by the IMRD structure will help keep you grounded and focused on the research.

Exercise 5B: If you are a student or an otherwise inexperienced STEM writer, jot down for yourself some external reasons for

- why you want to learn the IMRD structure, and
- why you want to succeed at IMRD writing about research in which you are engaged.

If these reasons are on our list just above, describe their specific relevance to the research you want to write about and how they may influence your writing.

For example, will the need to impress an instructor influence your style in the report? Will it influence how many data points you report in the results section?

Are there other external reasons for your wanting to write well in the IMRD form? Why are they important to you?

Writing for Your Intended Readers

You will note from the above section ("Considering Purposes") that it's not easy to write or talk about the purposes for your writing without also thinking about the readers you will be trying to influence by your research and writing. Indeed, all the external motivations involve positively influencing other people—especially those who can affect either your own advancement or career aspirations, or who can affect in some way how your field of research is regarded and respected. Actually, there is no reason why you should want to make a sharp separation between *why* you are writing and *who* you imagine reading your work and perhaps acting on what they read there.

PEER REVIEW AND THE IMRD STRUCTURE

After all, the IMRD structure itself came about and has thrived because of the communication needs of a vibrant community of researchers—some highly experienced, some finding their way, and some brand new—who are continually asking the kinds of questions of themselves and of other researchers that are implied by the IMRD structure.

We can see this with recent changes in some journals. Following what is sometimes called the replication crisis in the sciences, we have seen journals taking on a new approach to peer review and research articles. These are in the form of pre-registered studies where you only submit the introduction and the methods sections for review. In this model, you are setting up a study using the IMRD model, but before data collection or analysis sending out your study design for peer review. This helps move the demands of the IMRD structure from a focus on originality or novelty to strong study design.[1] What is important to take away is that this change is not merely one of process but also shapes the purpose of the research article genre.

1 You can read more about this important shift in the research article genre in Mehlenbacher (2019a).

Whether IM or IMRD, the nurturing idea of peer review gives life and continuity to the scientific enterprise, and peer review is guided by the questions that are basic to the IMRD form.

So, just who are your intended readers? Are they your fellow researchers in your area of inquiry? Are they people with a more general interest in and knowledge of your discipline? Are they science generalists? How you answer this question and how the editors of the journals in which you'd like to publish your research answer this question will determine how you write the article.

The editors of specialized journals in any STEM research area usually assume that their readers are professionally interested in that area of research. They will try to choose **peer reviewers** who are up to date in the research literature of that area of inquiry and who regularly publish in similar journals. So as a writer, you can pretty safely assume that you can use in your article

- the technical wording,
- the most common references,
- the known methods of analysis, and
- the known tests and calculations

of that research area. These assumptions about **language and context** are the major reason why research articles in specialized journals are often virtually unintelligible to readers from other STEM disciplines—not to mention undergraduate students and the broader readerships who require translations provided by journalists and other skilled generalist science writers.

Nevertheless, as we explored in Chapter Four, if researchers use the commonly accepted technical jargon in the specialized research area, then it is more likely that fellow researchers in *that specialty* will understand the reasoning in the article and will more quickly accept the results and significances of the research. Hence it is often more important to writers and editors to be convincing and clear to research specialists than to be intelligible to a broader audience.

However, not all readers of the specialized literature have the same level of technical language proficiency. It is crucial that writers (and editors) realize that even specialized peer-reviewed journals have some breadth of readerships in the community of the discipline.

For example, in our teaching experience with upper-level undergraduate STEM majors in the research university, these very intelligent students very often have great difficulty understanding highly technical articles in their own disciplines. Why? Because these students are still developing the technical vocabulary and advanced mathematics that can be assumed of PhD students and professional scientists in the specific discipline.

Moreover, if these students are still developing the tools to comprehend advanced research language, they will not yet be able to evaluate the methods and reasoning of the published writers. These critical skills, along with a specialized vocabulary, take years to develop. A major purpose of this book is to help developing students learn critical questions and methods they can apply to their reading and writing—in order to help them develop knowledge in their discipline.

So, does this developmental curve mean that writers and editors of specialized journals should seek to pitch their writing to a broader readership in their disciplinary community? That's a decision that editors and reviewers will need to make, and that they can communicate to their potential writers. They can do so in two ways:

- in their posted **author guidelines**—which writers should always consult before preparing an article for a journal; and
- in their **comments to writers** who have submitted drafts; frequently, journal editors will include in their forms for reviewers a question such as "Is this manuscript suitable for readers who" (for example) "also conduct research in this specialty" or "are broadly knowledgeable in this discipline, but who may not conduct research on these topics" or "conduct STEM research, but not necessarily in this discipline." Comments from reviewers and editors to writers about the nature of the intended audience can do much to help writers adjust their language to the target expectations.

Writers must always study the journal they seek to publish in to determine the level of language and knowledge of the research field it *seems* to prefer. Writers should always assume that there will be differences from journal to journal, editor to editor, regarding their sense of the breadth of their intended audience—even if these differences are not spelled out in the journal's author guidelines (which they rarely are).

Read through articles in a journal in which you'd like to publish and look closely for some of the following features:

- Diversity of the research topics in issues of the journal (including any special issues)

  Generally speaking, the broader the range of topics and research areas included in typical issues, the broader and less specialized the audience you will be addressing. What this means in terms of your writing practice is that the less specialized the intended audience, the more you may need to explain concepts and language.

- The level of technical language and assumed knowledge in the abstracts of the articles

  From the very first sentence of an abstract, for example, you may be able to tell just how much background knowledge of the research topic is assumed. As you read the abstract, notice how methods and results are described. What level of technical terminology and symbolic language is used—or avoided? Are technical terms defined, or is knowledge of them assumed?

- The amount of background description—such as definition of key terms, references to research history, descriptions of basic concepts—in the introduction

  Again, look for the level of technical terminology and symbolic language used. Are terms explained, or is it assumed that readers already know them?

- The level of technical language and assumed knowledge in the methods and materials and the results sections

  If these levels are consistent with levels in the abstract and introduction, then that consistency is a good indicator of the journal's intended audience. However, if methods and materials and results seem to assume a significantly greater specialized knowledge in readers, the editors may be projecting a bifurcation in the audiences for the different sections of the IMRD structure. (See more on this audience splitting below.)

- Lastly, how broadly or narrowly the discussion section applies the results of the research

  A narrow application of the significance of the results will be one that stays within, say, the methods of research within the discipline or in how key terms are defined by disciplinary specialists. This narrow application will more than likely indicate a highly specialized audience. Conversely, if the discussion moves to applications beyond the discipline, say to possible consequences for the larger society, then that may indicate a significantly broader audience that writers should expect.

AUDIENCE SPLITTING

As STEM writing, like all twenty-first-century communication, moves away from the print paradigm, writers and readers should expect to move digitally among related documents that were written for quite different audiences. We've already marked this movement in several chapters:

- In Chapter One, where a link in a piece of STEM journalism (on the relocation of endangered birds) takes us to the source research article in *Biological Conservation.*
- In the same example, the abstract of the research article is written with little technical jargon and with obvious awareness of a journalistic readership.
- In Chapter Three, where video gathering of data by animal researchers is readily available to millions via YouTube and project websites
- In Chapter Four, where more and more scientists are at ease moving between genres and audiences as different as those for TikTok, for project websites, and for their specialist colleagues in the peer-reviewed literature.

In all these examples, online linking enables readers of differing levels of technical knowledge to come into contact with writing meant for more or less specialized readers. Consider how rare this would have been prior to

- online publication and accessibility, and
- linking of journalistic summaries of research to the source materials.

Similarly, easy accessibility of video files of research sites has brought millions of viewers into contact with data gathering. This blurring of what used to be the sharp distinctions between "public" and "scientific" readerships has made both scientific specialists and much wider non-specialist audiences aware of the "other" as audiences with different expectations for communication.

Audience Splitting within the Individual Article

As STEM communication advances, we can expect to see even more audience splitting. Indeed, look for it within individual articles:

- Does the abstract seem pitched to a broader audience than the rest of the article?
- Does the introduction seem pitched to a broader audience than, say, the methods and results sections?
- For you as a STEM writer, this potential for audience splitting means an even greater incentive to becoming able to diversify the **terminology, style, and purposes** of your writing.

Exercise 5C: Choose a peer-reviewed article from a specialized journal in your field of interest. Read all five sections of the article: abstract, introduction, methods and materials, results, and discussion (this last section may have different names in different journals; e.g., conclusion, implications).

As you read, be aware of any differences in your sense of the audiences for each section. Use the lists of questions on pp. 126 and 127 of this chapter to guide your analysis. How would you characterize the different audiences that you perceive?

Do you also see some difference in the purposes of these sections that seem to be intended for different readers? How would you name these different purposes?

Now, study the author guidelines for the journal (usually linked from the main page of the journal website or located in the first few pages of the print journal). Is there any indication in these guidelines as to the journal's intended readership? If so, how is this stated? If not, write hypothetical wording for such a "guideline" that you think would be appropriate, based on your analysis of articles in the journal.

II. GIVING MOMENTUM TO YOUR RESEARCH "STORY"

THE SUSPENSEFUL MOMENTUM OF THE IMRD ARTICLE: PAST TO PRESENT TO FUTURE

We usually don't think of research articles grabbing the reader's attention the way a good story does. Indeed, readers often skip around in research articles, not reading them in their entirety but rather focusing on the sections they feel are most pertinent to their interests. And, frankly, some articles often encourage skipping around because the writers have not cared to make the article sufficiently coherent or interesting to encourage a thorough reading.

However, there's no reason why research writing has to put the reader to sleep—or confuse readers so they give up in frustration. The same excitement you feel when you are observing growth or change in the lab or in your field site you can communicate to your readers. Why? Because the IMRD structure is set up to do just that—if used by conscientious writers.

Think of IMRD as a narrative that moves from past to present to future. Not only that, but IMRD, like a suspense novel or a good movie, **first sets up problems or dilemmas to be solved**, then **leads the audience through a process**—sometimes straightforward, but more often twisting and turning—until **a solution or resolution is reached**.

Sure, there is a valuable difference between research writing and other kinds of good stories—science is never meant to tie up all the loose ends in a tidy ending. Good science always leaves us with questions to be answered—maybe even more questions than when we started. But what makes STEM writing valuable is that the process of the work—and how clear our reasoning is—takes the research community just a little further toward understanding something, and maybe even helps people and the world we live in a little bit, too.

Let's see how the IMRD structure can do that.

First, the introduction. The first goal of this section is to **focus our attention on a problem**: perhaps a question that people have been trying to answer, but unsuccessfully. The problem may be a troubling situation that has vexed people for ages or perhaps has just arisen in our advanced but still far-from-perfect civilization. Look at any research article introduction: how far into it do you go before some kind of problem is presented? Something that the writer(s) are trying to get you interested in, that they want you to care about so that you'll read on. (Do they make you care? If they don't, how could they?)

The introduction should also deepen your appreciation for the problem by showing how others have tried to solve it. Some writers treat this "review of the research" merely as a way to "check off" everyone who has ever published anything remotely related to the problem—as if the goal is to show us that they've read everything (see Chapter Three, p. 73, for more on this tendency, especially in inexperienced writers). However, writers can use the review of literature in the introduction to heighten our concern for the problem, if writers focus their attention on the prior research that has been most fruitful or original so far—or that perhaps has led science in a wrong direction that needs to be corrected.

In this way, the introduction can move the reader between the past and the present, even as it heightens our interest in the problem and how to solve it.

Lastly, the introduction presents us with the here-and-now of this research. The writers ask us to enter into their minds as they grapple with the problem: as they try to formulate the right questions, as they try to find the best setting in which to tackle the problem, and as they consider tools (e.g., technologies, tests, etc.) to help them—and the best, most appropriate people to assist them, either as co-researchers or as informants or subjects.

If the writers have been thinking of their readers' interest as they write, they are trying to make their readers as interested in the quest as they are.

Second, methods and materials. This section also moves the reader between past and present, as the writers carefully **describe what they did to set up the test(s), the tools they employed, the people they recruited, and why they did all these things**. The methods section is its own narrative within the larger story, because often this section goes step by step. And, again, if the writers want to keep their readers with them on this journey, they tell readers honestly how they decided upon which path to conduct the study and note the difficulties they encountered along that path. In Chapter Four, we stressed the ethical responsibility of researchers to be honest about the limitations of their research methods—about gaps between what they hoped to be able to achieve and what their tools or procedures might have limited them to. Part of the honesty of science includes describing difficulties and false starts and do-overs and adjustments, not only for honesty's sake, but, more important, to help other researchers.

Third, results. This section brings the story almost to the present, and it teases the reader with data that offer the promise of an answer to the problem that motivates the story. As described in Chapter Four, the results section should never be thought of as a "data dump," with every data point reported with no concern for organization and priorities. Rather, the results section is a necessarily selective, thoroughly honest **reporting of the data and the trends within** that relate closely to the questions that motivated the methods. (Chapter Six goes into more details on the results section.)

Fourth, the discussion. Here is where the story of the research comes full circle, as it brings us back to the problem and the associated questions with which we began the journey. However, much has happened along the way: the writers have

- designed a study by which they hope to solve the problem,
- carried out that study, as described in methods and materials, along with troubles and limitations they may have encountered, and
- reported to us, in results, the key data that have come from careful study.

Now, they tell us if the **questions have been answered and the problem solved**—at least temporarily!

But, more than that, now they also take us into the future, one where they project the possible significances of the research. Looking beyond this study, they consider what next might be done, and looking perhaps beyond that, they consider further research that may come about once the results of this research have been understood and applied. No, the research is never finished, but we now view the terrain from slightly higher ground, hoping for further breakthroughs. (See Chapter Six for much more on ways to write the discussion section.)

Exercise 5D: Read again the article you chose for the previous exercise on audiences (p. 130). This time, consider this article as a "story" that has the potential to engage the reader in the quest or journey of discovery that the article describes.

Apply our description of the stages to the story that the IMRD structure follows. What do the writers do in each section of the article to draw in the reader (you)? Do they miss opportunities to engage you? How might you change the article to make it more engaging?

Does the article seem to you honest in considering any limitations of the research or in reporting any false starts or do-overs that happened during the process?

Does the article in its discussion bring you back to the original problem and questions? Does it take you into the future? How? Are you left feeling that anything important has been left out? What would you like to see added?

Now that we have considered the rhetorical and story structures of the IMRD research article, let's begin to look at the individual sections. First, the abstract.

III. WRITING THE ABSTRACT

Think of the abstract as playing two roles in the research article. The first role is the one we all know: the abstract is a **short summary of an entire article**. We read the abstract in order to decide if we want to read the article itself. Often, if we are doing a scan of the research literature on a particular topic, we will read *only* abstracts in order to get a sense of the field. Well-written abstracts should give us a worthwhile sense of the "story" of the article, from introduction to discussion. (More on this purpose of the abstract below.)

The second role of the abstract is less well known but can be important to you as you write the article. Drafting the abstract can be a highly useful **tool for planning the entire article**. Because the abstract is meant to be a summary of the article in its entirety, a planning draft of the abstract can be an outline of what you hope will emerge in the article. In other words, the abstract draft is the writer's working tool.

Indeed, as your article moves toward completion, you could draft several iterations of the abstract to help you organize and refine your thinking and to accommodate what you are learning as you write the article.

Let's look at each of these purposes of the abstract, exploring the second one first.

THE ABSTRACT DRAFT AS PLANNING TOOL

The most common model of the abstract follows the IMRD structure. That is to say,

- the first sentence or two summarize your introduction,
- the next sentence or two summarize the methods and materials,
- the next sentence or two highlight key results, and
- and the final sentence or two summarize your discussion.

This is a handy way to think of the abstract, and it makes the tool ideal in planning your article. Moreover, the abstract should be a coherent summary, which means that each section should follow logically from the previous one and should, in total, provide a **coherent synopsis of the story of the article**. Thus,

- First sentences: What problem is our research addressing? What do we want to demonstrate?
- Second sentences: What methods and materials are we using to carry out this study?
- Third sentences: What are the key findings and types of data produced by our methods?
- Last sentences: Do the results demonstrate what we had hoped? What is their significance?

In planning your research article, you should be able to answer these four groups of questions succinctly. If you have a research project underway, try out this drafting tool to help you clarify your thinking and begin to put your ideas into words.

Drafting the abstract will help you see where your thinking is clear or still muddy, and where the various parts of your research may still not be coming together coherently.

Use abstract drafting to help you with best word choice. The drafting will also help you in the painstaking process of choosing the best words to capture your thinking for your intended reader. The tool is working well for you if you find yourself trying out different words.

Use abstract drafting to help you get good feedback. Take your draft of the abstract to a trusted reader—perhaps another member of your research team—and ask for feedback on the clarity, accuracy, and coherence of the draft. (Follow the advice on obtaining good feedback in Chapters Two and Four.)

THE FORMAL ABSTRACT

The abstract that precedes a peer-reviewed research article serves the vital role of helping readers decide if they wish to read the article itself. As a summary of the sections of the article, the abstract also helps readers decide which sections they might be most interested in reading, thus helping them focus their attention.

The abstract for a research article serves a similar function as an "executive summary" in business writing. The presence—and popularity—of both genres has come about because business people, like scientists and those interested in science, are extremely busy people. A well-written abstract or executive summary should create enough interest that a reader will spend some of their precious time with *this* article or report, rather than move on to another document that looks more interesting or meaningful.

In either case, the reader is always asking, "What's in this article or report for me? Is it important enough for me to spend some of my precious time reading further?"

Exercise 5E: Read several abstracts of articles in a field of interest to you. Compare them. Do the abstracts help you decide which article to read first, or which to read at all?

If so, what stands out in the abstract to attract your interest and curiosity? Conversely, is there an abstract that seems particularly unattractive? Why?

SIMILARITIES AND DIFFERENCES AMONG ABSTRACTS

Although the great majority of abstracts follow the template described on p. 134, abstracts differ, sometimes greatly, in

- length,
- format (e.g., use of subheadings and paragraphing),
- use of specialized technical jargon, and
- conformity to the template.

For example, use the four factors above to compare the following three abstracts (A to C) from different journals:

A. Relativistic Deflection of Background Starlight Measures the Mass of a Nearby White Dwarf Star

Gravitational deflection of starlight around the sun during the 1919 total solar eclipse provided measurements that confirmed Einstein's theory of general relativity. We have used the Hubble Space Telescope to measure the analogous process of astrometric microlensing caused by a nearby star, the white dwarf Stein 2051B. As Stein 2051B passed closely in front of a background star, the background star's position was deflected. Measurement of this deflection at multiple epochs allowed us to determine the mass of Stein 2051B—the sixth nearest white dwarf to the Sun—as 0.675 +/- 0.051 solar masses. The mass determination provides confirmation of the physics of degenerate matter and lends support to white dwarf evolutionary theory. (Sahu et al. 2017)

B. Shaping Planetary Nebulae with Jets in Inclined Triple Stellar Systems

We conduct three-dimensional hydrodynamical simulations of two opposite jets launched obliquely to the orbital plane around an asymptotic giant branch (AGB) star and within its dense wind, and demonstrate the formation of a 'messy' planetary nebula (PN), namely, a PN lacking any type of symmetry (highly irregular). In building the initial conditions we assume that a tight binary system orbits the AGB star, and that the orbital plane of the tight binary system is inclined to the orbital plane of the binary system and the AGB star (the triple system plane). We further assume that the accreted mass on to the tight binary system forms an accretion disk around one of the stars, and that the plane of the disk is tilted to the orbital plane of the triple system. The highly asymmetrical and filamentary structures that we obtain support the notion that messy PNe might be shaped by triple stellar systems. (Akashi and Soker 2017)

C. Integrating Phylogeography and Morphometrics to Assess Conservation Merits and Inform Conservation Strategies for an Endangered Subspecies of a Common Bird Species

Highlights

- Evolution of *Lichenostomus melanops* is explored with genetic and morphologic data.
- Genetic data suggests subdivision of *melanops* + *meltoni* vs *gippslandicus* + *cassidix*.
- IMa2 suggests *gippslandicus* and *cassidix* diverged in Mid-Pleistocene-Holocene.
- *Cassidix* harbors unique morphology and genetic diversity worthy of conservation.
- Introducing gene flow from *gippslandicus* may improve *cassidix*'s adaptive potential.

Abstract

Understanding the evolutionary history of threatened populations can improve their conservation management. Re-establishment of past but recent gene flow could re-invigorate threatened populations and replenish genetic diversity, necessary for population persistence. One of the four nominal subspecies of the common yellow-tufted honeyeater, *Lichenostomus melanops cassidix*, is critically endangered despite substantial conservation efforts over 55 years. Using a combination of morphometric, genetic and modeling approaches we tested for its evolutionary distinctiveness and conservation merit. We confirmed that *cassidix* has at least one morphometric distinction. It also differs genetically from the other subspecies in allele frequencies but not phylogenetically, implying that its evolution was recent. Modeling historical distribution supported a lack of vicariance and suggested possible gene flow among subspecies at least since the late Pleistocene. Multi-locus coalescent analyses indicated that *cassidix* diverged from its common ancestor with neighboring subspecies *gippslandicus* sometime from the mid-Pleistocene to the Holocene, and that it has the smallest historical effective population size of all subspecies. It appears that *cassidix* diverged from its ancestor with *gippslandicus* through a combination of drift and local selection. From patterns of genetic subdivision on two spatial scales

> and morphological variation, we concluded that *cassidix*, *gippslandicus* and (*melanops* + *meltoni*) are diagnosable as subspecies. Low genetic diversity and effective population size of *cassidix* may translate to low genetic fitness and evolutionary potential; accordingly we recommend managed gene flow from *gippslandicus* (with which it can be inferred to have recently experienced natural gene flow) to favor recovery of *cassidix*. (Pavlova et al. 2014)

Among differences that you'll perceive are the following:

- Length: Abstracts A and B are roughly 100 words in length, whereas Abstract C is about 300 words long.
- Format (e.g., use of subheadings and paragraphing): Abstracts A and B are presented as one paragraph with no subheads, whereas Abstract C is divided into a "Highlights" section, plus the lengthy abstract itself, all in one paragraph, which goes into detail about the history of divergence of the species focused on in the study.
- Specialized Technical Jargon: Abstract A employs some technical jargon and use of numerical data, and opens with sentences that are clearly accessible to a broadly non-specialist audience. In sharp contrast, Abstract B dives right into a description of methods and gives only one clause that might serve as a hypothesis: "[to] demonstrate the formation of a 'messy' planetary nebula (PN), namely, a PN lacking any type of symmetry (highly irregular)" (Akashi and Soker 2017). This clause is meant only for disciplinary specialists. Abstract C uses a bulleted list of "highlights" that will be more accessible to a specialist audience than to non-specialists (such as journalists). The abstract that follows includes a mix of accessible language and technical terms only known to those in the research field. It concludes with a justification of the procedure used by the researchers.
- Conformity to the Template: Abstract A clearly conforms to the IMRD template. Abstract B mostly describes methods, including "assumptions" on which they are based, with only the final sentence serving as a summary of discussion. Abstract C's "highlights" begin with a statement about methods, but then concentrates on results and their significance. The "abstract" of C covers background information usually found in an article's introduction, but moves to a statement of significance usually found in the discussion.

The comparison reinforces the value of the IMRD template in constructing the formal abstract.

The obvious differences among the three samples reinforce the advice for you to read closely the abstracts in the journals in which you would like to place articles.

Exercise 5F: Compare three or more abstracts from research articles in your field of interest. Use the four categories just above to look for similarities and differences.

Which of the abstracts seems to you the best at attracting reader interest in the article? Which seems to you the most in need of improvement?

If you have the chance to do this comparison with others in a group or class, listen to one another's responses.

Try revising the most deficient abstract to bring it up to the standard set by the best one. If possible, discuss your revision with others also doing this exercise.

Exercise 5G: Write a planning draft of an abstract for a research article that you are either writing or would like to write. Follow the outline on p. 134 that makes use of the IMRD structure. How does the planning draft help you think about the ideas, audience, and structure of the article that you are composing?

Be sure to write this draft so that it tells a coherent story about the research, while following the IMRD template.

Alternatively, if you would prefer to compose for your article an abstract that follows a different structure, such as that of the sample Abstract B above—or even that of sample Abstract C—try it out. What do you learn from this exercise about the ideas and language you want to use in order to reach your audience? Again, be sure to write this draft so that it tells a coherent story about the research.

With whatever option you choose for this draft abstract, discuss what you write with others also doing this exercise (if you have the opportunity).

IV. WRITING THE INTRODUCTION OF THE FULL ARTICLE

Let's review the questions that the introduction tries to answer (as described on p. 122 of the chapter):

- What is this research about?
- Why do you do it?
- What relevant research came before?
- How is this project different?
- What are you trying to show?
- What research questions are you trying to answer?

Now let's look in a bit more detail at each question. How might you try to answer it in terms of your own research interests and a study you are either part of or would like to undertake?

(**Suggestion**: Choose a research article in an area of interest to you. Keep it before you and study the introduction section as we look at each of the following questions. In some journals, the introduction section might be called "Background" or something similar.)

1. What is this research about?

Here is where you define the **topic** of your research as specifically as you can—and with language that your intended reader can understand. Depending on the knowledge level of your reader, you may need to go into some of the history of interest in this topic, so that your reader can appreciate the importance and value of research on this topic.

2. Why do you do it?

Here you define for your reader as specifically as you can the **problem or situation that has sparked the need** for your research on this topic. This is the internal motivation for the research project. Again, depending on your reader's level of knowledge about this research, you may need to include **evidence** of the importance of the problem that gave rise to your research.

3. What relevant research came before?

We emphasize the word *relevant* in order to argue against the tendency of some researchers to list in citations every article that perhaps in some way might have been conducted in related fields. If your article will appear in an online publication that links to the cited articles, you don't want to waste your readers' time by sending them to articles that have little or no bearing on your precise topic. In contrast, **citing** and in some cases **summarizing the most relevant prior research** will be important to you in **demonstrating why your research project is needed** (see Chapter Four for advice on ethical citing and summarizing).

Citing the most relevant research shows your reader how your research will be different from what has been done previously. It will also show how your research might build on a foundation established by predecessors—or perhaps correct what you see as inaccurate methods or conclusions in some prior research.

4. How is this project different?

This vital section of the introduction follows from your citation of the most important prior research. Here you can impress your readers by **defining as specifically as you can what makes your approach unique**—how it will build on or differ from the relevant work that has been done previously. You will do this by describing the proposed study, not by telling your reader how unique your project is.

5. What are you trying to show?

This is the first place in the article where you state your study's **goal or hypothesis**. This does *not* mean that you must state a specific outcome that you hope to achieve. Depending on the discipline and the particular inquiry, you *might* be able to express the goal in terms of specific outcomes. However, much research across disciplines is valuable for asking a new question or using a different methodology or applying methods to new materials or in new settings—and then analyzing the data closely to see what results emerge.

Being specific and vivid in defining "what you are trying to show" will help your reader focus intently on the sections of the article to follow.

6. What research questions are you trying to answer?

As with the previous bullet point, this question (about questions!) should be answered so that your readers will have something very firm to hold on to—and come back to—as the article proceeds. The reader will expect you to **come back to these research questions in your results section and, most emphatically, in your discussion**.

Some articles will not have lists of research questions per se. For example, in the article you are reading as you consider these bullet points, how does the writing team indicate the factors they will be looking at closely as they collect and analyze data?

> **Exercise 5H:** If you have been following the suggestion at the start of this section to read an article along with your reading of this section, write down the data you have observed. After considering your observations, what conclusions might you draw about how the article's introduction addresses these key questions? What revisions might you suggest?

SETTING THE STAGE IN THE INTRODUCTION—HOW MUCH HISTORY? HOW MUCH CONTEXT?

As noted above, a major purpose of the introduction is to build a case for the value of the research being described in the article. Often, building this case will require the writer to provide some history of the problem being addressed and how researchers have tried to address it. We might also call this effort to "build a case" an effort to provide a context for the research. When we try to answer the question "Why do you do this research?," we are providing this context.

However, writers must always deal with the dilemma of just how much history and context to provide. After all, virtually any research area will have a significant amount of history that could be repeated in your article. But the question is: how much of this history do you need to provide for these readers in this article?

The answer to the "how much?" question depends on three factors:

- the level of knowledge by the intended readers,
- the difficulty of the case you are trying to build, and
- constraint in the form of words allowed for the article.

The "words allowed" constraint may be the most basic factor of the three. Although proportions vary somewhat from journal to journal and article to article, keep in mind that the introduction usually comprises no more than one fourth of the total wordage for the entire IMRD article—and usually occupies much less, as readers are usually more interested in methods, results, and discussion than in the introduction.

So, even if you would like to go into more depth regarding the fascinating history of this research area, do the math and limit the introduction accordingly. Keep in mind, also, that the history portion makes up only a small portion of the introduction's concerns—so you may be limited to no more than a sentence (or a few) to set up the historical context. This means that you have room to describe only the most crucial historical elements that carry weight in your context-building for the research.

Given these basic logistical limitations, which historical elements you choose to mention will need to be based on your intended readers' level of knowledge and why your case may be difficult to build. For example, if your intended readers are highly experienced in the same type of research you are doing, you can assume that they know most of the history, so you don't need to demonstrate your knowledge of history to them. However, you might very well need to mention the past or recent event that most immediately impacts the focus and shape of the research that you are reporting in this article.

On the other hand, if your intended readers are not familiar with this research strand, you will still not likely have room for an extended history lesson—and they might not be interested in reading it anyway. But do consider if there are events or social circumstances they might already be familiar with and which might have some bearing on the purposes for your research—even if those events and circumstances are outside your field and might actually be more familiar to your audience than to you as a researcher. Reminding these readers of such a signal event could be a vital way of helping them see the importance of your research—in a way that they will readily understand. But *caution*: be sure that your research really is relevant to whatever event or circumstance you describe. As emphasized in Chapter Four, do not be guilty of "overclaiming" the value and applicability of your research.

Finally, the difficulty of the case you are trying to build for the value of your research may be the wild card in determining how much space to give to recounting history in your introduction. If, for example, your research goes significantly against the grain of the research in your area of inquiry, you may need to mention, even to specialists in your research field, events or theories with which they are not familiar, but that are indeed highly relevant to what you are attempting to do. Despite the fact that most IMRD articles devote relatively little wordage to the history portion of the introduction, use your judgment in determining if this specific justification demands a higher portion.

As always, in all these questions that depend on several factors, do not hesitate to call on your trusted readers to give you feedback on your ideas and drafts.

DETERMINING HOW MUCH CITATION YOU NEED IN THE INTRODUCTION

As with history, citing sources is part of building a context for your research. Within the introductions to articles that report new research, the number of citations is limited by the same factor that limits exploration of history: the relatively small proportion of the entire article given over to the introduction. Although, once again, journals and articles vary in number of citations, in general the introduction is not the place to indulge in citing every conceivable publication that might have been utilized in related fields.

Follow the same advice (just above) in terms of mentioning historical precedent: **cite only those sources that have clearly the most bearing on the research** you are undertaking. Be guided by the practice in the journals to which you wish to submit your article, and choose based on the needs of your intended readers and on your need to justify the value of your research. In general, less is more in terms of how much to cite.

If you are publishing online, it is especially important to be judicious and focused in your citations. The common practice in online journals of linking to source articles means that you are wasting readers' time if you are encouraging them to read articles that are not pertinent to your study. Again, be careful to note the citation practices of the journals to which you wish to submit.

Keep in mind, also, that another increasingly common online practice—linking your article to a supplemental section on methods and materials—can mean another opportunity to cite sources that are particularly important to your methodology. (See the section on methods and materials later in this chapter.)

How to Cite

As described in Chapter Four, tradition in peer-reviewed STEM journals frowns on direct quotation from source articles. (This STEM tradition is *not* followed in non-STEM disciplines, such as history, literature, writing studies, and politics, where direct quotation is often highly important to methodology.)

The STEM tradition means that writers are restricted to either

- mere listing of source numbers, or
- simple paraphrases of important sources.

Ethical paraphrase means that you must ensure the **accuracy of the meaning** of your paraphrase of the source. It also means that your paraphrase **may not include exact wording** from the cited source (see Chapter

Four, p. 102, on this issue). Because of these restrictions, paraphrases of sources tend to be very brief and emphasize the main reason for the pertinence of the source.

Again, because online publications often link to the sources cited, there is no reason for a writer to include a lengthy description of the source. As you read through introductions in peer-reviewed journals, pay attention to the ways you observe writers briefly describing sources they cite.

Citation styles. The great majority of peer-reviewed STEM journals use some version of the citation style of the **Council of Science Editors (CSE)**. One large STEM discipline in the United States, Psychology, uses the quite different style of the **American Psychological Association (APA)**. The author guidelines in the journals in which you wish to publish often include the journal's policy on citation styles. Be sure to read these guidelines carefully. Equally important, browse enough articles in that journal to get a sense of the version of the citation style that journal uses.

V. WRITING THE METHODS AND MATERIALS SECTION OF THE FULL ARTICLE

For many readers of STEM research articles, the methods and materials section (M/M) may be the most important—but, ironically, the most neglected—section of the full IMRD article. Why is this so? The reason for relative neglect is that most readers of research articles are more interested in the **findings and conclusions** of research, and so may not even read the M/M section.

Almost all readers of STEM articles will begin by reading the abstract, and most of those unfamiliar or only somewhat familiar with the research area will move to the introduction and then to the discussion, because the **subject of the research** (explained in the introduction) and the **significance of the results** (explained in the discussion) capture the interest of most readers.

Consequently, the results and M/M sections tend to be read closely by those with a practitioner's interest in this strand of research. Moreover, the M/M section's vital importance to the production of those results means that in many cases the M/M section is the longest and most intricately detailed section of the article—which means that it will go unread by all who are not deeply concerned with the methods themselves. That increasingly much of M/M is now *not* included in full in online publications, but is downloadable as a supplement, reinforces the tendency of most readers to ignore this vital section.

The true importance of the M/M section is often felt only when those who disagree with the conclusions reached by the researchers scrutinize how the study was conducted. Indeed, some of the most attentive readers of M/M are those who may feel they have a reason to dispute findings. It is not surprising that when disagreements are expressed, they are most often in terms of the methods used.

RHETORICAL CONSIDERATIONS OF THE METHODS AND MATERIALS SECTION

Purposes

The purposes of M/M are indicated by the **nine research questions** that the section tries to answer (as noted on p. 122):

- How did you design the research? Why this way?
- Where did you conduct it? Why?
- What steps did you follow? Why?
- What tools/participants/tests did you employ? Why?
- How were results recorded and the data secured?
- How were the data analyzed and significance calculated?
- If there are human participants, how did you ensure their safety and privacy?
- Show me everything you did, so that I (or someone else) could replicate your research or adapt it.
- Were there steps you wanted to perform, but couldn't? Why not?

We've set up all these questions—of which there are more for this section than for any other in the IMRD format—as if they are being asked by a skeptical reader who is reading closely.

You'll note that most of the questions also ask "Why?" or "Why not?" This indicates that the M/M section should illustrate **what you did** and **your justification** for those actions. It is rarely enough just to provide a "recipe" of the steps you followed and the tools you used.

As you read through M/M sections in articles in your area of research, keep careful track of how research teams

- describe the steps in the research and the tools employed,
- the reasons they give for these choices, and
- the sources they cite.

Answering the nine questions above (plus the whys and why nots) fully and accurately forms the internal motivation for M/M. The external motivations for M/M are

- to **convince** the careful and conscientious reader of the **thoroughness, accuracy**, and **appropriateness** of the methods and materials you employ, and
- to enable other researchers to **replicate** or to **adapt** your methods to their own work.

Intended Readers

So, who are these close and careful readers of your M/M? As noted above, they will most likely be only those with a clear professional interest in how your research was conducted, including

- editors of journals to whom you might submit the article,
- peer reviewers for those journals,
- principal investigators and fellow research team members for your research projects,
- potential funders and their expert reviewers,
- fellow researchers in the same line of inquiry or in closely related areas, and
- skeptics who may have reasons (professional, political, etc.) to dispute your conclusions.

Your intended readers of your M/M are not likely to include those with a casual interest in your research area. Therefore, you need *not* be concerned to make your M/M accessible to such readers in terms of its technical language. Focus on the language that you are confident that specialists in your research area know.

In contrast, as described earlier, the audiences for the abstract, introduction, and discussion may be much broader and less specialized, since interest in these sections is likely to be more widespread. For all these sections, if and how much to define or explain technical language will be a concern: one that writers will need the help of editors to resolve.

Once again, as you are planning your M/M section, read through M/M sections in other articles in your target journals to see how **technical terms** and **specialized symbolic language** are handled.

ORGANIZATION AND "READER-FRIENDLINESS" ISSUES IN M/M SECTIONS

M/M sections are likely to be longer and more detailed than any other section of the IMRD article (with the possible exception of results, as described in Chapter Six). Hence writers need to take particular care to achieve clear organization of these sections. It is not uncommon for the linked (or downloadable) M/M supplements to be organized, in fact, as if they were books or lengthy formal reports, with

- clear divisions and subdivisions by topic,
- bolded headings and subheadings, and
- a Table of Contents for ease of browsing.

In the Article Text or in the Supplement?

If the journal where you seek to publish tends to use these M/M supplements, you will also need to choose carefully which and how much of your methods you include in the text of the article itself and which information is found only in the supplement. Keep in mind that even specialist readers will tend to read the M/M information in the article text first, before going to the supplement—if they read the supplement at all.

Therefore, what you include of M/M in the text of the article itself might be what most of your readers will see. So, what do you choose to include? First, check the author guidelines for the journal you are targeting. In addition to explicit guidance about the M/M section, some journals have templates of the section and subsection titles to help guide you.

If your target journal does not have these features, then consider using the nine research questions (p. 122) as a template for designing a thorough M/M section. You might not wish to use these research questions as an outline for your section, but answering those questions will ensure that you have covered important topics that your readers will expect you to report on.

Ease of Reading

The ultimate outline you choose for your M/M section in the article text should be designed for ease of reading:

- Choose **headings** that conform to topics that your readers will most likely be interested in.
- Under each heading, do not hesitate to **divide sections into multiple paragraphs**, each of which is short enough to allow readers to focus and to help them not miss important information.
- **Remember**: Reader attention tends to drift in the middle of long paragraphs, especially in technical descriptions. Use **reader focus tools** such as paragraphing, bullet points, boldface type, etc. (see Chapter Twelve for more on these visual marking techniques) to help readers stay attentive. Again, check author guidelines for advice on these devices in your target journals.

Exercise 5I: Compare the M/M sections of three articles in a research area of particular interest to you. These articles should come from different journals, if possible. Observe and record the similarities and differences in the design and organization of the M/M sections.

Apply the nine research questions to each article. Are all questions answered? What appears to be missing?

Does each article use headings to subdivide the M/M section and heighten reader attention? Do any of the articles do this particularly well? How? What might be improved in any of the articles in terms of design and reader-friendliness?

Read the author guidelines for the journals. What guidance do they include for writers in terms of design and topics in M/M? Do you have suggestions for additional guidance?

CONCLUSION

This chapter has emphasized the vital importance of the peer-reviewed STEM journal article in the scientific process. IMRD (introduction, methods and materials, results, discussion) not only names the structure of this vital genre, but captures the process through which researchers describe their research focus and its inspiration. IMRD shows the research community how the writers carry out and assess their research, then demonstrates the significance of their findings.

Systematically, IMRD tells the careful, convincing, and often quite vivid story of research—and in so doing furthers the overall scientific enterprise.

We've scrutinized in this chapter three of the five sections of the IMRD structure—abstract, introduction, and methods and materials. In particular, we've investigated how writers can achieve the internal and external purposes of these sections—and how they can reach the ever-expanding and diversifying audiences for STEM research. The chapter offers advice on meeting these challenges. It provides exercises that make STEM students and professionals more savvy readers of STEM research articles, so that they can use what they learn through careful reading to make their own writing more accurate, thoughtful, and effective.

In Chapter Six, we rely on the IMRD foundation laid in Chapter Five to investigate the final two sections of the structure—results and discussion—which move the story of scientific process from the present into the future.

CHAPTER SIX

WRITING THE RESEARCH ARTICLE, PART TWO: RESULTS AND DISCUSSION

Topics in this chapter:

I. Results and Discussion in the Interconnected, Multimedia World
II. Distinguishing between the Results and Discussion Sections
III. Writing Results
IV. Writing the Discussion

Chapter Five described the overall structural and rhetorical elements of the IMRD-based research article, with advice on writing the first three sections: the abstract, the introduction, and the methods and materials. Here, we go into the final two sections of the IMRD article: the results and the discussion.

It may be best if you regard Chapters Five and Six as one. This chapter will refer occasionally to ideas and topics explored in Chapter Five, but *read that chapter before delving into this one*, because much of what we write here will depend on concepts first applied there. It will also help if you have practiced some of the exercises in that chapter before you explore the ideas here.

I. RESULTS AND DISCUSSION IN THE INTERCONNECTED, MULTIMEDIA WORLD

As with the abstract, introduction, and methods and materials, how the results and discussion sections are designed and written today vary greatly from what pertained in the print-only era. Readers now expect results in online articles to include

- a variety of **graphic elements**, such as colors, photographs, tables and graphs of many kinds, and sometimes video (when appropriate to the subject), and
- **links** to datasets; to supplemental, further-detailed explanations of results; and to sources.

Even print expectations have increased, as readers expect to find in print versions the URLs to the sites that are linked in the online versions or that can be downloaded as PDF supplements.

BE PREPARED TO MEET THE DIFFERING EXPECTATIONS OF TODAY'S MORE DIVERSE AUDIENCES

Moreover, as we saw in Chapter Five, online "linkability" between IMRD peer-reviewed articles and non-specialist sources (e.g., STEM journalism) means that you should expect a more diverse readership for all STEM research writing. This diverse accessibility means that editors of peer-reviewed articles in online publications may expect writers to address a broader, less-specialized audience. While this greater breadth of readership might have a lesser effect on how results are written (more on this later in the chapter), it might have a considerable effect on how discussion sections should be written.

Because of this greater diversity of potential readers,

- STEM writers need to carefully read the author guidelines in the journals they'd like to publish in to see if there are comments about reader expectations. If there are such comments, then writers should pay close attention to these instructions. But even if there are no such comments about an expected audience, **writers should not make assumptions about the audience**; instead,
- STEM writers need to read published articles in those journals closely to see what **level of technical specialization the editors assume about their readers** (see Chapter Five, pp. 125–28, for advice on what to look for in IMRD articles in terms of analyzing reader expectations).

Readers with different levels of interest and knowledge will have somewhat different expectations of the significance of the research results, and thus how discussion sections are written might be affected:

- **Specialist readers** might expect discussions of significance to be restricted to effects within disciplinary issues (e.g., preferred methods, meanings of technical terms, nuances of theory).

- A more **popular readership** may not be interested in such discussions. Instead, they may look for significance to be measured in relation to issues that affect the broader society (e.g., new applications and products, social policy, health care, economics). During the COVID-19 pandemic we saw, however, that non-expert readers may also be interested in the technical aspects of research discussions, and thus we must be mindful of such readers.
- **Journalists** are another audience to consider, as they may be reporting on the findings of a new study, and such readers may have some expertise in science, as many science journalists do, while others may not have training in science and report on a wide range of topics.

As we look at specific discussion sections later in the chapter, we will be examining ways in which writers handle this growing diversity of readers.

BE PREPARED TO INCREASE YOUR TECHNICAL PROFICIENCY IN HANDLING NEWER INFORMATIONAL TOOLS

As STEM writing adjusts to meet the changing demands of communication in a digital environment, expect that how you present results will continue to become **increasingly graphically sophisticated** as tools develop further. In addition, the merging of online publication with the ubiquitous cultural shift to social media means that the IMRD article can no longer be thought of as a "stand alone" artifact. Instead, the article is increasingly seen as just one important component within an intricately networked series of "conversations" at any given moment among hundreds or thousands, even millions, of interested readers and writers. Most of these readers will *not* read the IMRD articles: they will read newspaper and magazine accounts, websites of companies and news organizations, social media posts, and blogs. Still, many writers of these non-IMRD sources will have read the IMRD articles.

The IMRD article will continue to have great influence because of the power inherent in the **careful scientific method** and because of the **authority generated by peer review**. But the IMRD article's communicative power will work within a network of other documents that will be of perhaps equal power, including

- structured online commentaries by readers,
- blogs, microblogs, websites,
- videos, infographics, and other visuals,
- STEM journalism, and
- related mass media such as television, film, and games.

As a STEM writer, you will therefore need to prepare yourself technically—as well as rhetorically—to participate in this multi-genre, multimodal communication environment. Helping to initiate you into this adaptive process is a main purpose of this book.

The Paradigm Shift from Print to Multimedia

Whereas at one time online publication of research was frowned on by academics as inferior to print publication, now online publication has moved so far beyond print in its capabilities that the two have become increasingly dissimilar. When journal editors showed the research community that peer review was just as strong in online IMRD publications as in print publications, any reason for discrimination against digital publication fell away.

The print paradigm still influences the way in which some online articles appear—when, for example, an online journal strives to look like its bland, black-and-cream print counterpart. But the influence today is more often working in the opposite direction, as print versions either steadily disappear or strive to look more like the colorful, more varied screen interface (which drives up print costs and thus makes print journals even less sustainable).

Nevertheless, a caution: Print documents—and the structures, such as libraries, needed to hold them—still need to flourish, for several vital reasons:

- because print records are needed as a backup system not dependent on the relatively fragile (e.g., hackable and dependent on the power grid) electronic infrastructure,
- because print books and other paper artifacts are still a portable technology that millions of readers prefer for some of their reading, and
- because so much of the past, including in STEM fields, has not been recorded digitally and is not likely to be.

However, STEM writers today should not be guided by print's limitations in planning how to present results or think about the discussions of their research. Rather, in planning the tools you need to communicate your research, consider

- the technologies and tools that you and your colleagues are most likely to use to communicate with one another,
- the technologies and tools that you and your colleagues are most likely to use to communicate with other readers important to you and vital to the influence of your research,

- the tools and media you use for learning in areas of interest to you, and
- the IMRD articles that are most convincing to you, and how results and discussion are presented in those articles.

STUDYING A SAMPLE IMRD ARTICLE

Let's look at a recent IMRD article to see how the new online paradigm plays out in the presentation of results and discussion.

> Controlled Hydroxyapatite Biomineralization in an ~810 Million-Year-Old Unicellular Eukaryote
>
> **Abstract**
> Biomineralization marks one of the most significant evolutionary milestones among the Eukarya, but its roots in the fossil record remain obscure. We report crystallographic and geochemical evidence for controlled eukaryotic biomineralization in Neoproterozoic scale microfossils from the Fifteenmile Group of Yukon, Canada. High-resolution transmission electron microscopy reveals that the microfossils are constructed of a hierarchically organized interwoven network of fibrous hydroxyapatite crystals each elongated along the [001] direction, indicating biological control over microstructural crystallization. New Re-Os geochronological data from organic-rich shale directly below the fossil-bearing limestone constrain their age to $<810.7 \pm 6.3$ million years ago. Mineralogical and geochemical variations from these sedimentary rocks indicate that dynamic global marine redox conditions, enhanced by local restriction, may have led to an increase in dissolved phosphate in pore and bottom waters of the Fifteenmile basin and facilitated the necessary geochemical conditions for the advent of calcium phosphate biomineralization. (Cohen et al. 2017)

This article from *Science Advances* demonstrates several of the factors of online IMRD articles described above. For example, the abstract is written for an audience minimally familiar with the concept of biomineralization, but it does not assume knowledge of its history or significance in dating the origins of life of one-celled creatures:

> Biomineralization marks one of the most significant evolutionary milestones among the Eukarya, but its roots in the fossil record remain obscure. (Cohen et al. 2017)

Note the accessibility of this sentence to readers with little background in the type of research reported here.

Similarly, the introduction shows balance between assumed **technical vocabulary** from the research discipline (e.g., "coccolithophores," "Neoproterozoic") and **explanations accessible** to a less-specialized science reader:

> Combining these microanalytical results with a new radiometric age constraint on the fossil assemblage establishes the ASMs as the oldest known eukaryotic representatives of biologically controlled mineralization. (Cohen et al. 2017)

While the writers use the acronym "ASM" throughout, it is initially explained as "apatitic scale microfossils." Fellow researchers already know this information, but non-specialists do not, and so this explanation is written for them.

In contrast, the results section (which in this article, and often in this journal, comes right after the introduction) assumes the reader's knowledge of crystalline structures and analysis methods common in the research area:

> Selected area electron diffraction analyses acquired throughout the specimens further indicate that the fibrous HAP crystals are everywhere elongated along the [001] direction. (Cohen et al. 2017)

That the abstract and introduction are written for a broader audience than is the results section exemplifies the **audience splitting** described in Chapter Five (pp. 129–30). Even more emphatically, the methods and materials section demonstrates a high degree of reader familiarity with methods and tools common in the research area, and so the methods and materials section is written for a much more specialized audience than are the abstract, the introduction, and, as we shall see later, the discussion section.

Graphic presentation in the article depends on the reader's access to digital tools (although overall graphic usage in this article is minimal in comparison to graphic usage in other articles). To read the one graph and two infographics (one drawn and one photographic), a reader must be able greatly to enlarge the images or to download them. For example, the photographic infographic, "Electron micrographs of an ASM specimen," details features observed with electron microscopy and then highlights even smaller features of each micrograph. Without appropriate digital tools, a reader would not be able to observe these features.

The article links readers to downloadable supplementary materials in PDF. These supplements include, for example, "all geochemical and geochronological data presented in the paper" (Cohen et al. 2017).

Exercise 6A: Choose an online IMRD research article from a journal that you read and in which you might like to publish. Read the entire article, but focus on the results and discussion sections. Observe the following characteristics:

- Identify technical knowledge and vocabulary that the article assumes readers will know (and that therefore are not explained) in (1) results and (2) discussion. Make a short list.
- Does there seem to be a difference between the amount of assumed knowledge in the results and the amount of assumed knowledge in the discussion? Is there evidence of audience splitting (described in Chapter Five) between these two sections? If so, cite examples.
- Characterize the technologies and specific tools used to present information graphically in the results section. With which of these tools are you familiar from your own work? Which would you need to learn in order to use? Is there one tool that stands out to you as particularly useful or interesting for you to learn?

II. DISTINGUISHING BETWEEN THE RESULTS AND DISCUSSION SECTIONS

The IMRD format implies a sharp distinction between results and discussion. However, in practice, it is not easy for writers to choose which information and analysis to include in the results and which to include in the discussion. This difficulty arises because *both* sections require

- a reasoned **analysis** of data, and
- statements about the **significance** of what has been observed.

For example, consider these sentences from the results section of the biomineralization article:

> The primary microstructure preserved in the ASMs reflects a high degree of biological control over HAP crystallization. Morphological control over skeletal components, specifically the development of fibrous HAP elongated along [001], can only be achieved through the inhibition of selected crystal faces during growth. (Cohen et al. 2017)

The first sentence makes a **conclusive statement** about what has been observed: "reflects a high degree of biological control." The second sentence illustrates the conclusion of the first sentence and asserts that this control "can only be achieved through the inhibition of selected crystal faces during growth." These **analytical statements of significance** would not be out of place in the discussion section. That they are in the results section illustrates the difficulty of determining what goes into one section vs. another.

(Indeed, in some journals and for some articles, editors have chosen to combine results and discussion in a single section. We will not be analyzing such combined sections in this chapter, but you should be alert to this possibility in the journals you read.)

How the two sections differ. If we look again at the questions generated about each IMRD section at the beginning of Chapter Five, we can understand more clearly the reason for this potential confusion—but also begin to see a way out of the impasse:

- **Results**—What were your findings (or what have they been so far)? What have you selected to show us in this article and why? What forms do those results take—calculations, data points, graphed trends, transcripts, images, etc.? What data are you merely summarizing? Why? What types of data are being left out of this section? Why?
- **Discussion**—What do these findings mean to you? Do they answer your initial questions? Why and how? Or why not? What other kinds of significance might your results have? How confident are you about the significance you are claiming? How have you measured that confidence? What limitations or cautions should we acknowledge about what you have found? What are the next steps for this research? Why? What still needs to be done?

The distinction that is being implied in these two sets of reader questions is fairly straightforward. The results section answers the **reader's desire to be shown the most important data**. In order to meet this desire, the researcher-writer must first analyze the data in order to select

- what to show,
- how to display it, and
- what to say about it.

In order to select what to show, the researcher-writer must reach conclusions about the significance of all that has been observed, and then must convey to the reader the reasons for that selection. This necessary analytical process toward selecting what to show and how to show it creates the confusion with the discussion section, and this confusion can't be avoided.

Nevertheless, what stands out about the results section is its focus on **showing and analyzing data**. For example, the analytical sentences quoted from the biomineralization article are appropriate in results because they focus the reader's attention on the significance of the arrangement of the crystals observed in the research. Moreover, note that it is in the results section that we learn about the electron microscopy of the crystals, graphically reinforcing the focus on the data.

In contrast, the questions regarding the discussion section push the researcher-writer to think about significance differently. The questions reflect an expansion of the context of the analysis from the immediate data observed in the research site to

- much larger **spaces**,
- a greater span of **time**, and
- issues whose **relevance extends beyond** the current study.

So, for example, notice that the discussion section in the biomineralization article is subdivided into three subtopics that signal this great **expansion of context**:

- "Geochronological Implications and Context"
- "Conditions at Mount Slipper"
- "Paleobiological Implications" (Cohen et al. 2017)

The first and third of these headings clearly indicate the expansions of context listed above in terms of space, time, and relevance. The second of these headings, "Conditions at Mount Slipper," appears to narrow the geographic focus to the research site for the study. However, the argument of the subsection is just the opposite: that the "local conditions" at the research site (Mount Slipper, in the Yukon, Canada) indicate a much larger process affecting even "modern" events "across continental margins" in general:

> Phosphate deposits are neither ubiquitous nor the most important products of redox oscillation in *modern marine basins*. Instead, major episodes of phosphogenesis, largely postdating the Fifteenmile succession, speak most directly to the interaction between pore water PO_4 and Fe, microbial ecology, and the concentration of phosphorus at the seafloor in response to *pervasive oxidation across continental margins*. (Cohen et al. 2017; emphasis added)

The context continues to expand as the discussion section draws to its close. The first sentence of its final paragraph signals this expansion of relevance:

> These results contribute to an emerging and nuanced picture of early Neoproterozoic *tectonic, environmental, and biological change.* (Cohen et al. 2017; emphasis added)

The "Paleobiological Implications" become explicitly speculative:

> Our data leave *open the possibility* that the proliferation of eukaryotes and the advent of phosphate biomineralization were connected by a fundamental reorganization of the P cycle. Although *this relationship is speculative* on the basis of our current understanding of Tonian biogeochemical cycling, our data do support an increase in bioavailable phosphate during this interval of the Neoproterozoic. (Cohen et al. 2017; emphasis added)

The final sentence of the section ends with a speculative possibility that stresses the potential historic impact of this localized study in the Yukon:

> The enhanced bioavailability of P provided a new, if temporally limited, opportunity for eukaryotes to explore biomineralization for what may have been *the first time in Earth's history.* (Cohen et al. 2017; emphasis added)

RESULTS VS. DISCUSSION—DIFFERENCES IN FOCUS AND SCOPE

Therefore, the results and discussion sections **share an analytical purpose that leads to statements about significance.** This shared purpose often confuses writers about what is appropriate to include in each section.

However, the two sections consider significance differently in both focus and scope.

The results section focuses on the **collected data in the study**. Analysis of significance is limited to

- the research site(s) and data collected there, and
- trends and patterns in the data.

The discussion section expands focus from the local site and methods of the study to a **much broader context** that may include implications for

- many more places and situations,
- a much greater span of time, and
- relevance to more and larger issues.

The importance of this difference for you as a writer is that you can use it to build reader attention and interest as you move toward the conclusion of your article. We will see how this can happen as we consider the next two sections of the chapter.

Exercise 6B: Choose two IMRD articles in a research area of interest to you. Read both and pay closest attention to the results and discussion sections of each. Compare the two articles for the following features:

- In the results section of each article, on which data and analysis do the researchers-writers concentrate?
- What statements of significance do the writers make about these data?
- Do there appear to be important data left out of the analysis?
- In the discussion section of each article, how do the researchers-writers expand the context of the article away from the data collection site(s) to introduce larger issues and perhaps more places and a larger time span?
- What statements of significance does each article make in the discussion?
- In how broad a context of space, time, and issues are these statements made?

Bottom line: Which of the two articles do you feel makes a stronger case for its statements of significance? Why? Compare your views with those of others, if possible.

III. WRITING RESULTS

Writing your results section, as emphasized earlier in this chapter, requires you to make a range of decisions about your data and how you have analyzed its various components. Among these decisions are the following:

- Which data will you report in the results section—and how have you made those choices?
- How will you display the data?
- How will you explain your data selection to your readers?
- What statements of significance will you make about the data you are presenting?
- How will you justify to your readers these statements of significance?
- How will you organize the section (using subheads) so that your readers can follow your presentation of data and analysis?

Let us look at each of these decisions.

WHICH DATA WILL YOU REPORT IN THE RESULTS—AND HOW HAVE YOU MADE THOSE CHOICES?

Remember that readers will not want you to dump all your data on them and expect them to sort through it all to judge its significance. That's your job and that of your team members (if you are part of a team). Your goal in results will be to identify, for your reader, the **trends** and **patterns** in your data that relate most clearly to the focus of your study that you have announced in your introduction. Strive to be **concise**.

The data you choose to describe in your results will be those **that support the statements of significance** that you choose to make in the section. Remember also that online journals will enable you to add **supplements** (often in PDFs). The supplementary materials can include much more data than you will want to present in the results section of the article itself. Indeed, more and more journals will expect you to include links or URLs to enable readers to find these supplementary databases, should they wish to see them.

How you choose which data to present in results often depends on the statements of significance you will decide to make about the data. In other words, you will of course need to carefully and systematically pore over your data in order to decide what the data are telling you about the phenomena you have been studying. To do so honestly and thoroughly is one of the most critical parts of the scientific method, and it cannot be rushed.

Writing to learn (see Chapter Two) can be of immense help in systematically poring over the data and trying to make sense of it in a way that will enable you to report your statements of significance to your readers. Depending on the methods and the measurement and analysis tools you use, it can be immensely helpful to you to draft such items as lists of what seem to you the most important data and possible statements of significance you may want to make about the data. Think of your writing in this phase of your work as a flexible tool to help you reach understanding and clarity in your thinking—as well as to try out possible wording that you think your intended readers will understand. You might show your drafts to teammates for their feedback. Never hesitate to **revise** your writing with an aim to refining your message for your intended readers.

HOW WILL YOU DISPLAY THE DATA?

Look again at Chapter Three to consider the tools you might use to display your chosen data, and how these options might work to gain the attention of your readers and convey your information vividly and convincingly. As Chapter Three asserts, the digitally published journal offers a huge array of tools for presentation of your data and its analysis, so choosing what's best for your data and your statements of significance will be part of your challenge.

Keep in mind that, just as the results section is not a data dump, it should also *not* be thought of as an excuse to overwhelm your reader with as many charts, tables, photos, graphs, equations, etc., as you can muster. Be judicious. As researchers, we tend to want to show all our work, both because we're proud of it and because we feel we may need to justify the expenses that have been paid by a funding source. But always keep in mind your sense of what your intended readers will need to understand and stay focused on your argument in the article.

Today, there are also repositories where data can be stored and shared. These offer important resources for researchers and can help with replication studies and/or allow those who want to know more about the data to investigate it themselves. Sharing data also means that the study's findings can continue to be judiciously reported to ensure that readers who are looking for key insights are provided with them, while those who wish to know more can turn to the original data.

In choosing how to display your data, pay attention also to the **interplay between words and visual/audio media**. Readers' focus will almost always be drawn to a graphic, especially if it is colorful. The graphic should not only catch the reader's attention, but should also quickly convey a relatively simple message that can reinforce a statement of significance in the text. For example,

- a line graph should indicate a trend;
- a photo or drawing should clearly highlight features described in the text; and
- a table should also help the reader focus on the important trends in the data—it should *not* be so clogged with data points that a reader cannot

easily discern a pattern in the data—a pattern that the text should be able to point out without having to account for all the extraneous data points.

See Chapter Three for suggestions about how to make your charts and tables communicate, not confuse.

Rule of thumb. If you want to use a particular graphic, be sure that your text will not need to explain so much about the graphic that the reader loses your point. A series of rainbow-colored charts that present data at several stages of a process may look pretty, but if there are too many variables represented by those colors and too many values shown for each variable, it might be best to create a simpler chart that reinforces your statement of significance about the trends in the data.

Exercise 6C: Choose three IMRD articles in an area of interest to you and observe how data are displayed in the results. Have the writers been judicious in selecting graphics? How do the graphics reinforce the statements of significance in the text?

Which of the three articles is most effective in using graphic tools? Which is least effective? How would you suggest graphics be changed to increase their effectiveness?

HOW WILL YOU EXPLAIN YOUR DATA SELECTION TO YOUR READERS?

The results section will often include an introductory paragraph or so that summarizes

- the types of data that have been collected, and
- the methods (approaches, tests, etc.) you have used to analyze these different types of data.

These paragraphs may include information about methods that have already been described in the methods and materials section (see Chapter Five), but writers may choose to describe the data analysis tools *only* in results. As you read articles from different journals, you will see which of these two practices a particular journal tends to prefer.

Regardless of the options in describing data analysis methods, you will need to **account for the different types of data you have collected**, and how you have selected the data you wish to emphasize in the results section. You will want to assure readers that all relevant data have been carefully considered as part of the research process. If, for the purposes of a specific study, you will only be considering some of the data, you will need to explain the reasons for this selectivity to your readers.

It may well be, for example, that you have collected much more data of different kinds than are really applicable to the research focus in the article. Or it might be that the different collections or series of observations have produced data of different degrees of usefulness. Whatever the reason for data selection, they should be explained and accounted for. As described in Chapter Four, a vital aspect of the *ethos* of science is the reader's confidence in the integrity of the research process. Accounting for the data collected—and your reasons for selecting some but not all of the data for analysis—is part of building reader confidence.

Exercise 6D: Re-read the same three articles you chose for the previous exercise. This time, focus attention on those paragraphs of the results section that describe the type of data collected and the methods of data analysis. How thorough are these descriptions?

How do the sections address selectivity of data for analysis? Do the sections assure you that all relevant data have been carefully analyzed? Do questions about this remain for you in any of the articles? How would you rank these three articles in terms of their accounting for the selectivity of data analyzed?

WHAT STATEMENTS OF SIGNIFICANCE WILL YOU MAKE ABOUT THE DATA YOU ARE PRESENTING?

As you carefully analyze your data, you will arrive at an understanding of trends and patterns in the data. Identifying these trends and patterns for your readers will be your main goal in the results section and will focus your readers' attention. Moreover, making statements of significance about these trends and patterns will help you organize the results section. Lastly, by making these statements of significance in results, you will pave the way for the broader implications of your research that will be the focus of your discussion.

In order to arrive at the statements of significance you wish to make in results, we recommend following the writing-to-learn process described above (pp. 162–63). As you reflect on the data before you, including the measures you have used to analyze the data, keep this question in mind:

> What statements of significance can we (the research team) make with **confidence** about the **trends** and **patterns** we are seeing in the data?

Write tentative statements of significance as a way to clarify and refine your thinking—*but* be ready to modify these draft statements as your analysis of data continues. Through this reiterative process, you should arrive at significance statements that the data support, and that you can report to readers with a high level of confidence.

Thus, for example, we have the statements of significance we cited earlier (p. 157) in the results section from the biomineralization article:

> The primary microstructure preserved in the ASMs reflects *a high degree of biological control over HAP crystallization.* Morphological control over skeletal components, specifically the development of fibrous HAP elongated along [001], *can only be achieved through the inhibition of selected crystal faces during growth.* (Cohen et al. 2017; emphasis added)

Such statements focus the reader's attention. In doing so, however, they set up the expectation by readers that you will justify that high level of confidence. Therefore, once you have arrived at your statements of significance, your next task will be to justify those statements to your readers.

HOW WILL YOU JUSTIFY THESE STATEMENTS OF SIGNIFICANCE TO YOUR READERS?

In the results section of the biomineralization article, the first two paragraphs address the subtopic headed "Mineralogy of ASMs." After a first paragraph summarizes the tools and tests used to measure mineralogy, the second paragraph begins with the statements of significance highlighted above. It then conveys a systematic argument to justify the researchers' confidence in the statements of significance. The paragraph addresses the differences between non-biological crystallization and the "biological control" that the statements claim. Reference to sources is the primary method the researchers-writers use in the paragraph to reinforce their analysis of differences. The last sentence summarizes the three points they have made in the paragraph to defend their claim:

> *The* [1] *preservation of consistent hierarchical organization, the* [2] *persistence of HAP,* and *the* [3] *retention of preferred crystallographic orientation* indicate that ASMs were formed via a controlled eukaryotic

> biomineralization pathway and that their original structure was largely unmodified by diagenetic replacement or recrystallization. (Cohen et al. 2017; numerals and emphasis added)

By creating the paragraph in this systematic way, the researchers-writers address the reader's expectation that they will defend the statements of significance that they made at the start of the paragraph. The researchers-writers do not leave their readers waiting for a justification, but meet that reader expectation immediately.

Note, also, that this example demonstrates once again that the results section is by no means a presentation of unanalyzed data, but a carefully organized argument that

- makes statements of significance about selected data, then
- uses evidence to justify systematically the significance claims.

Exercise 6E: Using the same three articles you have used in the previous exercises, identify at least one statement of significance made in the results section of each article. How does each article justify the significance claims?

What types of evidence are used to support the argument? References to sources? Statistics? Equations? Images and charts, etc.? Worded descriptions of observations?

In your judgment, which of the three articles succeeds best in justifying claims? Why?

HOW WILL YOU ORGANIZE THE RESULTS SO THAT YOUR READERS CAN FOLLOW YOUR PRESENTATION OF DATA AND ANALYSIS?

We have now observed how researchers-writers (1) create statements of significance about the mass of data they have collected, (2) present the data that most pertain to these statements, and (3) argue, with evidence, to justify the statements of significance. However, we have not yet looked at the overall organization of the results section.

Basically, your overall results section should be designed to

- focus your readers' attention on those trends and patterns in the data that most clearly relate to the objectives of the research, as announced in your introduction, and

- pave the way for the larger statements of significance that you will make in your discussion.

You do *not* want your results section to cover topics in the data that are *not* relevant to the focus of this study. Introducing those non-relevant topics (no matter how interesting they are to you for other reasons) will only confuse your reader. Save those topics for other articles, or include them, if you see a reason to, in the supplements that you'll link to the article.

The biomineralization article we've been using as a sample in this chapter is a good example of a concise results section that limits coverage to those data topics clearly relevant to the aims of the study and to the larger statements of significance that will be made in the discussion to follow.

This results section consists of four lengthy paragraphs organized under three subheads:

- "Mineralogy of ASMs" (two paragraphs)
- "Geochronological Constraints on the ASM Assemblage" (one paragraph)
- "Mineralogical and Geochemical Analyses from the Fifteenmile Group" (one paragraph) (Cohen et al. 2017)

Each of these topics relates clearly to statements of significance in the data and to topics that will be important in the discussion. Moreover, they follow directly from the focus of the study announced in the introduction:

> Evidence of *pre-Ediacaran biologically controlled mineralization* has never been definitively identified. Thus, both the *timing and the environmental circumstances* surrounding the evolution of biologically controlled mineralization are unknown. (Cohen et al. 2017; emphasis added)

By using **subheads**, the researchers-writers signal to readers the focal topics and enable readers, if they wish, to move among topics depending on their interests. Most importantly, thanks to subheads, readers don't need to search through each paragraph to discover its focus.

Reinforcing the clarity provided by the subheads, each topic is designed as an argument that makes a statement of significance—followed by an evidence-based justification for the statement. We have already looked closely at the topic "Mineralogy of ASMs" (pp. 166–67) to see how it adheres to this pattern and enables the targeted readers to follow easily.

Exercise 6F: Using the same three articles you have used in Exercises 6C, 6D, and 6E, study the organization of the results section in each. In your judgment,

- Are the topics in the section clearly marked for ease of reading? How? With subheads, for example?
- Is each topic in the section relevant to the focus of the study announced in the introduction?
- Within each topic, do the paragraphs stay focused on the topic and clearly justify a statement of significance regarding data?

Which of the three articles seems most successful in creating an organization in the results section relevant to the study and is clearest to its intended readers? What improvements would you suggest?

IV. WRITING THE DISCUSSION

Think of your discussion as the end toward which each section of the IMRD article has been pointing:

- Your well-written abstract will likely include a final sentence or two that forecasts what will be asserted in the discussion (see Chapter Five, Section III).
- The introduction will focus the article on the issues and questions that the discussion will ultimately answer (see Chapter Five, Section IV).
- The methods and materials section will detail the approaches and tools that will be used to bring out tangible evidence about the focal issues (see Chapter Five, Section V).
- In a final lead-up to the discussion, results will show readers the most important data brought out by the methods and materials and will make statements of significance of the data that will prepare the way for the larger implications to be shared in the discussion (see Chapter Six, Section I).

If you have done your job of designing each section of the article, the intended readers will be hungry for the **broader implications** of your research.

Indeed, many readers of IMRD articles are so hungry for these implications that they jump from the introduction—or even from the abstract!—to the discussion section. These readers (and all of us are guilty of reading this way from time to time) can't wait to find out where the researchers ended up.

If reading the discussion then makes these readers curious about how the researchers got to their conclusions, they may then make time to read other sections. Ironically, as we pointed out in Chapter Five (p. 145), the section often relegated to last-read (if at all) is the one that most clearly embodies the scientific process: methods and materials.

But so be it. The point is that you must carefully craft your discussion section, because your readers will be interested.

TYPICAL SECTIONS OF THE DISCUSSION

The specific design of discussion sections will vary somewhat from journal to journal, and even from article to article. As you read the literature in your field, you can look for differences.

Nevertheless, when most readers come to the discussion, they are looking for

- **statements of significance**—often titled "implications"—that move beyond the statements made about data in the results. (If necessary, review the distinctions between results and discussion at p. 160.)
- **projections of what further research is needed**—what next steps are suggested by this study? What researchable questions are implied by the findings of this study?

Of these two expectations by readers, the first—implications—is by far the most important to readers. Indeed, many IMRD articles do not include the projections of further research. Again, as you read IMRD articles, you'll see how individual journals handle this option.

DESCRIBING IMPLICATIONS

The discussion section offers the researcher-writer some **freedom to imagine** the wider significance of a study than do other sections. As we noted in describing the biomineralization article, the results section is tied to finding verifiable trends and patterns in the data. The discussion section is not limited in this way.

Indeed, the idea of implications allows researchers to explore connections and analogies between what has been found in the research site and through the methods used there and other situations, times, and places.

But the researcher is still constrained by the need to provide a rationale—an argument—for the analogy or connection. For example, the biomineralization team makes a connection between crystal composition and formation in the current research site in the Yukon and marine/coastal deposits from many other times and places. The paragraph making this connection **argues for the relevance of the analogy** by citing other studies. Drawing an implication like this is far from being a flight of fancy or a whim, even though it does not offer the immediate verifiability of a data point in the current study.

THE LANGUAGE OF IMPLICATIONS: TAKING CARE IN SPECULATION

The freedom offered to researchers in the discussion means that both writers and readers of IMRD articles **need to look closely** at how analogies and connections are phrased. Readers must be aware of how writers are stating the analogy, so that they do not mistake a verifiable (or verified) statement of significance for a merely plausible speculation.

Likewise, writers—if they are to write ethically (see Chapter Four)—need to take great care in how they phrase possible connections. For example, when the biomineralization team speculates about the implications of its Yukon study, note the guarded language:

> Our data *leave open the possibility* that the proliferation of eukaryotes and the advent of phosphate biomineralization were connected by a fundamental reorganization of the P cycle. Although *this relationship is speculative* on the basis of our current understanding of Tonian biogeochemical cycling, our data do support an increase in bioavailable phosphate during this interval of the Neoproterozoic....
>
> The enhanced bioavailability of P provided a new, if temporally limited, opportunity for eukaryotes to explore biomineralization for what *may have been* the first time in Earth's history. (Cohen et al. 2017; emphasis added)

Rhetoricians call such guarded language "qualifying language" (or "qualifying terms") because it tries to capture the **quality** of, or **extent** of, **confidence** the writers have in the statement made. One responsibility of STEM writers is to make their qualifying language as **accurate as possible** in estimating their confidence in statements of significance.

Similarly, STEM readers have a responsibility to pay attention to qualifying language, so that they do not misread the source literature.

(**Note**: We will return to this topic of qualifying language in Chapter Thirteen, when we focus on cutting unneeded words, one of the fixes for common errors in STEM writing styles.)

Regardless of the level of qualifying language researchers-writers use in stating their implications, they will need to argue for the analogy using evidence from the results or external sources to achieve credibility—ethos—with their intended readers.

Exercise 6G: Again, using the three articles you've used for earlier exercises in this chapter, read carefully the discussion section of each. For each article, identify one or more implications, as defined here. Write down the implication you've chosen.

Then observe closely how the writers argue for that implication. What evidence do they cite: other studies? Data from the research? Other types of evidence?

Rank the three articles in how well they argue for their implications. What improvements might you suggest?

Exercise 6H: Now look closely again at the implications you chose from each of the three articles. This time, pay attention to the *qualifying language* the writers use in expressing implications in the discussion sections. Write down the qualifying words or phrases. (Some common terms include "seems," "may," "possibly," "probably," "sometimes," "might," and many other related words or phrases.)

Think about the use of such terms in stating these implications. Do they affect your reaction to the implications? Do they appear to express the appropriate level of confidence? What changes would you suggest?

CHOOSING IMPLICATIONS: CONSIDERING THE READER

The relative freedom that researchers have to choose implications to describe and justify in their discussions depends in part on the readers they are trying to influence. In other words, it's a **rhetorical consideration.**

In writing their discussion sections, all researchers-writers ask themselves and their team members, "How far from the immediate findings of this study should we go in expressing implications?" The answer is never obvious, but depends in part on the **intended readers** and on the team's **level of confidence** in any implication they might put forward.

If, for example, you see your intended readers as other researchers in your specific field of interest, then the implications you'll consider will concern issues that are well known in that narrow field, and you'll not be likely to venture into wider territory. For example, in the biomineralization article, the researchers were interested in two main research questions:

- if the eukaryotes on Mount Slipper demonstrated biological control of the crystallization patterns, and
- the date that biological control first occurred.

Both of these questions are well known and have been studied by other researchers in the field. In choosing implications to express, they could have restricted their statements of significance to the patterns of data they observed at the research site (and that they reported in results).

However, they chose to expand the context for their implications to marine/coastal sites across continents and to speculate whether this study showed such biological control for the "first time in Earth's history" (Cohen et al. 2017). While these are questions of interest in the research field, they are also questions of interest for many other groups of readers, including nonscientists. And so, by broadening the implications across space and time, the researchers are making their implications relevant to a much broader and diverse audience.

CHOOSING IMPLICATIONS: AN ETHICAL CONSIDERATION

As a rhetorical consideration, choosing **how far to expand the context** for implications of research also becomes an **ethical** consideration. In Chapter Four, we considered the danger of overclaiming the results of a research project (pp. 107–15). While making unjustifiable claims is more likely in such media as STEM journalism, blogs, social media, and lab websites, we also raised this concern for peer-reviewed IMRD articles in Chapter Four. We bring it up again now, because the discussion section is the **most likely place for overclaiming** to occur in the IMRD article. STEM researchers should always weigh

the desire to influence readers versus **the accurate level of confidence**

in deciding which implications to express in the discussion section. If those two factors are not **balanced**, and the desire to influence outweighs the justifiable level of confidence, the implication should be avoided.

Because the discussion section gives researchers the freedom to explore possible implications of their research, it is all the more important for researchers-writers to use that freedom ethically. That is why we have emphasized the need for writers to justify the implications they express—and why we have urged readers of STEM research to be particularly watchful for

- qualifying language, and
- the types of evidence used in arguments for implications.

Exercise 6I: Think about research in which you have been engaged or which you'd like to undertake. In writing about that research or even in describing it orally to others, what have you implied about its potential benefits or results?

In making these implications, have you been concerned with how difficult it might be to justify the implication? If so, you are weighing the desire to influence against the justifiable level of confidence. What actions might you have to take to balance these opposing factors? How might you perhaps modify your implications or improve your methods and tests to raise the level of confidence?

Write about this difficult choice. What have you learned?

CONCLUSION

This chapter has built on the foundation of Chapter Five with the aim of helping you

- write results and discussion sections in IMRD articles, and
- become a more discerning, questioning reader of STEM research.

Taken together, both chapters should help you craft IMRD articles that tell well-organized, well-focused, well-argued, and interesting stories that your intended readers will want to read.

If you do the exercises throughout the chapter, you will definitely be helped along this path.

In the next chapter, "Writing the Research Review," we will build on the lessons of Chapters Five and Six toward learning a vital genre of STEM research that has received much less attention than its important place in the progress of science has merited.

CHAPTER SEVEN

WRITING THE RESEARCH REVIEW

Topics in this chapter:

I. Goals of the Research Review and Comparison with the IMRD Article
II. Features and Types of Research Review
III. Rhetorical Considerations in Writing the Research Review
- Ethos (includes the Artificial Intelligence Challenge to Ethos)
- Purpose
- Audience (includes Audience Splitting)

I. GOALS OF THE RESEARCH REVIEW AND COMPARISON WITH THE IMRD ARTICLE

When junior and senior STEM majors have come into our courses in "Writing in Science," some of them are not familiar with the genre known as the **research review**—even though they may already have read some of these review articles in peer-reviewed journals that they have perused for other courses. Some of these STEM majors have trouble distinguishing a review from an IMRD article. This is understandable, because the review of research looks very much like the introduction section of the IMRD reports of new research. Research reviews also share some of the characteristics of the discussion sections of articles of new research.

Indeed, some of these reviews of research are even structured in IMRD format (see Section II, Type B below), even though they review the work of other scholars and do not in themselves present new experimental research.

Like the introduction of the IMRD article, the review article begins by defining a specific research topic or question. Then it proceeds to describe some of the **history of research** on that topic or question.

But here the similarity to the IMRD article mostly ends.

The IMRD reports of new research go on from the introduction to describe in great detail a single, new research study—its methods and materials, its results, and the implications of the new research in the discussion. The IMRD research article is the focus of Chapters Five and Six of this book.

In contrast, the review of research has a very different but related purpose. The review goes into depth to describe the history of research on this specific topic, and to bring that research path up to the present. Then it projects into the future to answer the question, "**Where should the research go from here?**"

Why are the writers of the review interested in this history and in projections toward the future? Because their larger goal is to **influence how we understand the implications** of this research progress and to **influence how this research will be carried on in the future**.

In this focus on the implications of the research and on future research, the review bears some similarity to the discussion section of the IMRD article. As we explored in Chapter Six, a major purpose of the discussion is to argue for implications of the research that often go considerably beyond the immediate results of the focal study.

DIFFERENCES BETWEEN THE RESEARCH REVIEW AND THE IMRD STUDY

1. The basic structural difference of the research review from the IMRD article is the review essay's *lack* of detailed sections on methods and materials and results. Because the research review is in most cases not focused on a specific study, but **serves as an overview of research progress** on a specific line of research, any description of methods and materials will likely focus on methods used to analyze the research of others. Any results described in the research review are likely to be either
    - a summary of results found by other researchers, or
    - a description of the results of the analysis of the research of others.
2. Because there are often (though not always) no detailed sections on methods and materials and results, which in many cases are the longest sections in the IMRD article, the research review is likely to be shorter than the IMRD experimental article. Though this is not always the case (see Section II of this chapter), research reviews frequently serve as relatively **quick-to-read summaries** of trends of research on specific lines

of inquiry; hence their brevity. When research reviews serve this summary focus, it is not uncommon for reviews to be only one to three pages (500–1,000 words). Further, just as the overall length of the research review is often much shorter than that of the IMRD experimental study, the **number of references is often far lower**, though this is not always the case (see Section II of this chapter).

3. As a summary of research progress in a specific line of inquiry, the research review is usually **intended for a broader audience** than would read a specific study in an IMRD article. Given the varied purposes of research reviews, audiences might vary from researchers with some practical knowledge of research in the specific area all the way to very broad, blended audiences of science professionals and non-scientists. (See Section III of this chapter, below, for more on rhetorical considerations, including audience.)

II. FEATURES AND TYPES OF RESEARCH REVIEW

The genre of the research review includes types that may be as concise as 500-to-1,000-word summaries of the latest developments in an area of scientific inquiry and as deep and detailed as entire books on an area of research. Works as different as these share the basic goals of the research review:

- to influence how we understand the implications of this research progress and
- to influence how this research will be carried on in the future.

Regardless of the length and complexity of the research review, the general structure of the genre can be summarized in a template:

Template of Six Common Features of the Research Review

- **history** (either summarized or in detail) **of interest in the research questions** that this research strives to answer
- **history** (either summarized or in detail) **of research progress** relevant to this line of inquiry, usually leading up to the "present time" (that is, when the review was written)
- **focus on one** (or a small group) **of the most important studies** (usually summarized either briefly or in detail)
- **description of the current state of theory** to which this research path has led—including any major differences of opinion that still remain

- **implications of this research progress and resulting theory**—including impacts on research in this inquiry or perhaps extending to larger social, political, or environmental issues
- **recommendations for next steps in research** relevant to this line of inquiry

Of course, how these features are displayed in any given journal or from article to article will vary. Nevertheless, as you think about writing a review of research in an area of interest to you, keep these features in mind as a template for your building of information and ideas for the review.

Moreover, as you read research reviews in peer-reviewed journals, apply this template in order to see how the writers have (or have not) followed this structure in making their arguments about implications and future research.

Also, as you read, note that journals will often call the research review by other names:

- essay
- feature essay
- perspective
- opinion
- summary of research
- research article

At times, our students have been confused by research reviews that they initially assume are IMRD experimental studies because the review has been included in a journal among IMRD articles, with no separate named category. Such reviews of research appear to be experimental studies, but are actually analyses of the research of other scholars. (See Section II, type B, below, for an example of such a review.)

Often, however, when students do not immediately see the difference between research reviews and new experimental studies, this confusion has been caused by their doing a **keyword search** (using a search engine such as Google or Yahoo, or in a library database) that brings up IMRD experimental articles and research review articles with similar titles.

Exercise 7A: Keeping in mind the basic goals and the template of features of the research review, consider how you might write a research review of a line of inquiry in which you are particularly interested. Use the template as a spark to guide you toward learning more about your research interest, its history, and its path of development. As you learn more, begin to fill in the outline of the template.

As you search for information, pay particular attention to any relevant research reviews that may have already been written in your area of research or in related areas. How do these research reviews fit within the template of features or vary from it?

THREE TYPES OF RESEARCH REVIEW

The description above of the general features of the research review covers many research reviews you will read. However, the genre is varied, and here we describe three different types of reviews that you will likely encounter—and may be interested in writing.

Type A: The Research Review Focused on a Specific Recent Study

One type of research review focuses on the implications of a specific recent study. The research review article places that new study in the context of research on a specific question.

For example, "The Genome—Seeing It Clearly Now" (Larson and Misteli 2017), published in *Science*, focuses on a new study by Ou et al. (2017; published in the same issue) that reports on the use of a different imaging technique, abbreviated "ChromEMT," to reveal the structure of DNA within a cell nucleus.

This 1,000-word research review sticks closely to the template of features described above. It opens by naming the research question that the new study (Ou et al. 2017) addresses:

> It is a curious fact that in many fields of science, central questions that one might assume to have been answered decades ago remain perennially unresolved. In genome biology, one such cornerstone problem has been how DNA is organized in an intact cell nucleus. (Larson and Misteli 2017)

The authors then zoom in on the importance of the new study: "Ou *et al.* now take a major step toward answering this foundational question" (Larson and Misteli 2017).

Staying with the template, the review then summarizes the **history of research** related to the question of DNA imaging, most of which has involved "biochemical in vitro studies," which have revealed what have come to be regarded as typical structures of chromatin fibers. The authors keep the review article firmly focused on the contribution of Ou et al. (2017) by asking

> But why do we not know what this compacted chromatin fiber looks like in a cell nucleus? (Larson and Misteli 2017)

This dramatic question sets the stage for the next section of the review, which describes in summary the difficulty of imaging DNA in vivo and the shortcomings to this task of electron microscopy (EM) methods—until the advent of ChromEMT.

The authors' summary of the history of research leads to their description of the factors that make ChromEMT more effective at showing the organization of DNA in chromatin fibers in the actual cell. Here the authors of the review, Larson and Misteli, following the template, briefly summarize the methods and the results of the Ou et al. (2017) research team, so that their **intended readers understand what the recent research has achieved and how it has been achieved.**

This summary of methods and results covers roughly the second third of the review article. As part of the summary, Larson and Misteli (2017) explore a second important question that the Ou et al. (2017) study addresses: "how different chromatin states are generated in a cell nucleus." This allows them to describe what the results show.

The next key stage of the research review template, the **current state of theory** on the research question and research questions that are sparked by the new research, is signaled by another dramatic question: "How sure can we be of these results?" This question moves the review into its final third. Here the authors describe several alternative explanations of the results and then speculate **how further research could help to answer them**.

The final paragraph of the review opens with a sentence that helps to achieve the second basic goal of the research review, to **set an agenda for further research**:

> The ability to visualize the chromatin fiber in the cell nucleus is a landmark achievement that opens the doors to probing chromatin structure in relation to its function. (Larson and Misteli 2017)

The remaining sentences give examples of this further research, before the review ends with a sentence that reinforces the impact of the Ou et al. (2017) study:

> These are challenges for the future—for now, one of the enduring foundational questions in the field of genome biology has been answered. (Larson and Misteli 2017)

The Larson and Misteli research review demonstrates clearly an **argumentative structure** that adheres to the template of features of the research review. Note once again the use of key sentences, including questions (as shown above), that maintain well-organized emphasis on

- the focal study, and
- the research questions identified in the opening section of the review.

Note also that the **argument** of the research review is supported by **evidence from prior research** that is cited by the authors and summarized. As in the discussion section of the IMRD article (see Chapter Six), the argumentative structure and the use of appropriate evidence in the research review enable the authors to state

- reasonable implications of the research results, and
- a reasonable agenda of future research.

Exercise 7B: In a journal of your choice, read a research review (remember that the name of this form may vary from journal to journal) that accords with one of your research interests. As you read, identify key sentences that

- focus the review and
- move the argument of the review forward.

Look closely, also, for the evidence that supports assertions in the argument.

Lastly, pay attention to the future research that is recommended. Do you find both evidence and research agenda appropriate and reasonable?

Type B: The Research Review Constructed in IMRD Format

The IMRD format is so pervasive in peer-reviewed journals that it is often used to **construct and present analytical reviews of the work of previous scholars**, rather than present new experimental research. These reviews are also often categorized in journals as "research articles" because they require laborious and meticulous research and analysis among many experimental studies conducted over many years.

Moreover, this Type B research review can make an important contribution to new knowledge if it analyzes prior data in a way that **leads to a new understanding of the line of research**. For example, see the sample Type B research review, below.

However, the type of analysis required is not that of the data collected in the laboratory or field experimentation, which most readers of STEM journals expect to read. Instead, it is **comparative analysis** of the methods and results of the studies by many previous scholars whose work the review examines. This is valuable research in its own right, as it contributes to advancing understanding of a line of inquiry—but it differs from the research carried out by most authors of STEM research articles presented in IMRD format.

It is easy to understand why students often confuse such reviews with experimental reports in STEM journals, because they follow the IMRD format.

As with the Type A research reviews described above, these Type B reviews have as their goal the same two objectives of all research reviews:

- to influence how intended readers understand the implications of this research progress, and
- to influence how this research will be carried out in the future.

A case in point is "Pesticide Residues and Bees—A Risk Assessment" (Sanchez-Bayo and Goka 2014). This research review, categorized in *PLOS One* as a "research article," announces its focus in the abstract:

> Using data from recent residue surveys and toxicity of pesticides to honey and bumble bees, a comprehensive evaluation of risks under current exposure conditions is presented here. (Sanchez-Bayo and Goka 2014)

Introduction. In their introduction, Sanchez-Bayo and Goka fulfill the first part of the review template by describing the **history of recent research** on the effects of pesticides on the decline of bee populations. They introduce the history as follows:

> Growing concern about the impact of pesticides on pollinators is reflected in the enormous literature on the topic in the past few years. In response to this concern, considerable amounts of new data on toxic effects of pesticides on wild bees, in particular bumble bees, have been obtained from laboratory and semi-field experiments.

> ...Naturally, low levels of pesticides may act as stressors that make bees more prone to biological infections. Among the pesticides, newly developed systemic insecticides such as fipronil and neonicotinoids have been targeted as the main culprits involved in the collapses since they were launched to the market in the mid-1990s. (Sanchez-Bayo and Goka 2014)

Later in the introduction, the authors home in on the more specific reason for their assessments of prior studies: the risks to bees caused by pesticide residues. They announce the goal of this review and the sources of their data:

> Here, we attempt to provide a comprehensive risk assessment for all pesticide residues found in pollen and honey, or nectar, to bumble and honey bees *using all residue and toxicity information available to date in the open literature and databases.* (Sanchez-Bayo and Goka 2014; emphasis added)

Methods and materials. Unlike the short Type A research review described earlier, this Type B review is much more comprehensive in terms of studies cited and includes an extensive methods and materials (M/M) section, which describes how the authors analyzed their long list of studies and measured the risk assessment.

To enhance clarity for intended readers, the authors use multiple sub-headings to organize the M/M section. These subsections focus on sources of data, agreement of data and methods among sources, and Sanchez-Bayo and Goka's methods of data analysis. Subheads are as follows:

- Residues Data
- Toxicity Data
- Data Analysis
  - Standard Risk Approach
  - Risk of Synergistic Mixtures
  - New Approaches to Risk

This expanded, well-organized M/M section is essential to the goals of this research review as it **presents new approaches** to risk assessment of pesticide residues that affect honeybees and bumble bees. Therefore, the use of the IMRD format for this Type B review article is fully appropriate to its goals.

Results. Because this research review presents in its M/M section new approaches to risk assessment in its line of inquiry, a detailed results section is necessary and appropriate.

In its list of subheads, the organizational structure of results follows from the emphases on the new approaches to risk assessment described in M/M:

- Residue Data
- Risk by Contact Exposure
- Risk by Dietary Exposure
    - Dietary Risk to Honeybees
    - Dietary Risk to Bumble Bees
- Risks by Cumulative Toxicity

Discussion and conclusions. The implications of the results in this review study follow from the emphases in both the M/M and results sections. It should be noted that the authors *don't* use subheads in the discussion, and so the organization of key ideas is less easy to find in this part of the article. Nevertheless, close reading of the 12-paragraph discussion shows that two types of implications stand out.

The first is **methodological**, focusing on the approaches to risk assessment used in this article:

> Traditional risk assessments have considered only the residue loads in pollen and the acute oral or contact toxicity of the compounds. We draw attention here to this important distinction, as the toxicity of hydrophobic insecticides and acaricides is mostly by contact exposure whereas the toxicity of hydrophilic fungicides and systemic insecticides is mainly by oral ingestion of residues in pollen and honey. (Sanchez-Bayo and Goka 2014)

The second is to identify the **insecticide residues posing the greatest risks** to honeybees and bumble bees:

> In view of these findings, banning of some neonicotinoids by the European Community seems to be justified alone on the grounds of residues in the food of bees, apart from other considerations and side-effects that these compounds may have. Surely, the high prevalence of neonicotinoids in honey (17–65%) is of great concern not only for worker bees but also for larvae. (Sanchez-Bayo and Goka 2014)

Recommendations for further research. In this example of a Type B research review, the only explicit recommendation for further research occurs in the final sentences of the Conclusion:

> In addition, risks of systemic neonicotinoids are probably underestimated because of their time-cumulative toxicity, synergistic effects with ergosterol inhibiting fungicides, and additive effects in combination with pyrethroids. *Further research on the combined effects of such mixtures is needed* to fully understand the reasons behind the collapse of honey bee and bumble bee colonies. (Sanchez-Bayo and Goka 2014; emphasis added)

Some explicit re-emphasis on the authors' new approaches to measurement of risk assessment would have been useful to readers at the very end of the article, in order to clarify the impact of this review article on future research.

In summary, this sample Type B research review demonstrates when the use of the IMRD format is fully appropriate for the genre of the research review. The extended M/M and results sections in the Type B review are essential to showing the contribution of this review to

- intended readers' understanding of the line of research, and
- priorities for further research.

Exercise 7C: Among IMRD research articles in your field of interest, find at least one that is in fact a Type B research review. As you read the article, ask yourself these questions:

- Is the IMRD format appropriate for this review?
- Do the methods and materials and results sections contribute meaningfully to the strength of the review? How?
- Could this research review be presented as effectively and usefully to readers in the Type A format? Might the Type A format be more useful? If so, how?

If you are doing this exercise in a class setting, compare your findings with those of others.

Type C: The Research Review as Book

In STEM writing, books hold an ambiguous place. There is no doubt that the IMRD research article, published in a peer-reviewed journal, is the most prestigious writing in STEM fields (though researchers might argue that successful grant proposals run at least a close second!). The IMRD research article is the avenue for advancement of scientific knowledge. The peer-reviewed journal is where scientists and science journalists alike look for what is new and most important.

When STEM academicians seek promotion, IMRD articles and other scientists' citation of those articles significantly determine advancement. It is common for STEM academics and researchers never to have written a book.

In contrast, some other academic fields, especially in the humanities, give pride of place to the book. It is the rare humanities scholar in a research-focused university who does not need to have had published a well-received book in order to advance in the institution. Indeed, article publication is far less important, and it is not uncommon for humanities departments to award tenure to scholars who present one published book and very few articles.

In those research fields, the book embodies work of several, often many, years. For younger scholars, the research begun for a PhD dissertation is characteristically carried forward into an academic appointment, with the goal being to publish the fruits of this long-term research—the book—before a sixth-year tenure decision.

Moreover, the book is typically seen as an **independent, one-person work**. The common STEM notion of the "research team" is thought strange in those environments. "Research" is conceived of as a mostly solitary activity, usually spent in archives and libraries, or in home or office intensely scrutinizing print documents or online versions of them, or in transcribing interviews, and so forth, depending on the discipline.

Then why devote any space in this STEM writing guide to the writing of books by STEM writers?

Here's one reason: the reading that humanists do for their book writing is divided between what humanities scholars call "primary" and "secondary" sources. Primary sources are those works that are the subject of the research—the plays of Shakespeare, the speeches of Nelson Mandela, the paintings of Georgia O'Keefe, etc. Secondary sources are all the commentary on these primary works by fellow artists, critics, journalists, and scholars. In short, the books written by humanities scholars are to a major extent *research reviews*—very like the Type B research reviews described in the previous section of this chapter.

Also like the STEM research review, the major goals of scholarly books—whether in STEM or other fields—are

- to influence how intended readers understand the implications of a line of research, and
- to influence how this research will be carried out in the future.

Is there a second reason to discuss book writing in this STEM writing guide? Yes: the book, unlike the Type A and Type B research reviews, provides **length** and **opportunities** for **varying style and tone**—as described in Chapter One (pp. 40–42)—and therefore opportunities for reaching audiences who would rarely, if ever, read an IMRD article on the subject of the book.

But varying style and tone also lets the writer/researcher **reach fellow scientists** in ways that the IMRD article does not allow. Some of these choices include

- using a **first-person point of view**, which allows for expressions of opinion, stories of personal experience, etc.,
- **varying point of view** within the work, from, say, a detailed description of a lab or field process to the writer's reflections on that work—or commentary by others about the author or about the subject of the work,
- **sustaining a critical argument**, which allows the writer to explore at length the opinions of others and to argue for an approach or interpretation based on multiple kinds of evidence, and
- using **quotations** from the writing of others (which the IMRD format does not allow) or **pieces of conversations** in order to enliven prose with human interest and precise language.

In short, book writing gives STEM writers the freedom to use—and develop skill in using—tools of writers from other fields and in other genres—even as the STEM writers strive to maintain the integrity of scientific thought and process.

As we go on to consider in coming chapters such genres as science journalism (Chapter Eight) and STEM blogs (Chapter Nine), we'll see other opportunities for STEM writers to branch out stylistically. Nevertheless, the book is a time-honored way by which STEM writers have reviewed research in their areas of interest in order to meet the basic goals of the research review.

STEM Books and You

Think of your own growth as a science student and working scientist, and consider how books about science and scientists may have played a role in your own inspiration and in your own choice of life's work. STEM books take many forms—for example, the many series of books for children of different ages, books usually filled with photos and drawings, often including recipes for simple experiments. Or biographies and autobiographies, which give us great insight into the lives, choices, and reflections of well-known researchers in many disciplines.

Think, then, of the many famous, even classic, books by scientists, books which have changed the way people think about the world, or at least which have introduced thousands of readers to the excitement of research and to ideas they'd never imagined. Most books by scientists have not been famous, but if they have reached even small audiences in shaping ways, they have done their job. Think of a book that you've read either in the past or recently that has had this kind of shaping effect on your own thinking. What are the traits of that book to which you've most responded? What tools of style and tone—such as those noted in the bullet points above—has this writer used to reach you and have that profound effect on you?

A book? Yes. But what about plays, films, and videos? What we've said so far about books also applies to works in other popular genres such as plays, films, and videos. Consider, for example, the long-running *Nature* and *Nova* series on PBS, which have inspired generations of young people to go into science and huge numbers of people to become interested in scientific issues. Though these programs are definitely examples of STEM journalism—the subject of the next chapter—the direct participation and leadership by scientists in these projects is essential to their success.

Exercise 7D: Recall a book (or perhaps a play, motion picture, or online video) about science that has been important to you either in the past or now. Write informally about why this work has been important to you, and identify what in it inspired you. What tools of style or tone or genre did the author use that you particularly remember—a personal story, a convincing argument, a description of a place, a conversation, a dramatic event, a pivotal experiment? Some other feature?

Follow up Exercise 7E: Think of your own research interests and experience. Write informally for a few minutes about an event or place or process from this research that has some meaning for you, and that you'd not like to forget. In your writing, practice using one or more of the tools of style or tone mentioned in this section of the chapter. Give yourself some freedom to experiment with your writing. (You might, for example, try using one of the techniques or approaches that most impressed you in the work you considered in the previous exercise.)

III. RHETORICAL CONSIDERATIONS IN WRITING THE RESEARCH REVIEW

Consider again all three types of the research review described above, from the short summary of research progress (Type A) through the extended IMRD review of many previous studies (Type B) to the varied styles of the book, play, movie, etc. (Type C). In all three types, the writer's attention to **purpose, audience**, and **ethos** is essential.

ETHOS IN THE RESEARCH REVIEW

Let's start with ethos. Throughout the book thus far, we've emphasized that the credibility of the scientist-writer comes from the commitment to the scientific method, as well as from the prestige accorded scientific publication by the peer-review process in STEM journals. (Chapter Four specifically goes into this point.) When STEM writers report their carefully observed and analyzed phenomena in IMRD articles, their well-chosen methods and plentiful results confer credibility on their work. Their judicious and reasonable implications bolster their standing as scientists. The several parts of the IMRD structure work together (see Chapter Five) to present a favorable picture to fellow researchers.

However, the research review presents a challenge in terms of ethos. Since the Type A review does not include methods and materials or results sections, how does the reviewer achieve credibility for the argument of the article regarding implications of the research and recommendations for future studies?

One way is by publishing the review in a reputable journal. For example, the sample Type A review we studied in section I, on DNA imaging, achieves ethos through its publication in *Science* (Larson and Misteli 2017). In addition, the review appears in the same issue with the IMRD study that reports this new method of imaging. In other words, this path to ethos **depends on the association of the review with the environment in which it is published** and the material that accompanies it.

So a less published or experienced STEM practitioner may rightfully ask, "How can I achieve sufficient credibility to publish a review?" This question is not really different from the more basic question of how any new scholar can break into publication. Chapter Two addresses this question for those writers, students especially, who are still grappling with the anxiety of putting words to paper or onto a screen—as well as the less basic concern of getting your writing "out there" so others can see it and weigh its value.

For those writers who have already reached the stage where they write regularly for others in STEM genres, the STEM community offers many opportunities for writers to build both their confidence and their credibility in writing. The fact that research is carried out in teams, and that team publication is the norm, gives new researchers/writers many avenues to contribute to publishable work—and to take on more and more responsibility for writing articles and reports. The team structure, when it works well, gives newer scholars the opportunity to learn from experienced scholar/writers, to get feedback on their work, and to be part of projects that reach publication.

So we'll concentrate now on the second way to achieve ethos for your research review: the **quality of your research and argumentation**. Remember that the research review (see pp. 175–77) uses a method of research that

- compares the theories put forward by other researchers, and
- argues for particular implications of the research thus far in this line of inquiry.

So the quality of the research in a research review depends on

- the **thoroughness of your reading and citing** of the most pertinent studies, and
- the **logical reasonableness of your argument** for a particular point of view.

Although it may seem as if this type of research is not a typical part of the research paradigm in STEM, in fact—fortunately—this method of research and reasoning is much like the one you would use for the introduction of the IMRD experimental article and for the discussion section of that article (as we covered in Chapters Five and Six). In the introduction, you are using the citation and analysis of earlier studies to justify your own research. In the discussion, you are using the results of your research to argue for implications that go beyond the immediate research setting.

Moreover, if you follow the template of features of the research review (pp. 177–78), you will have a structural organization for your argument that most readers will be able to follow, and that moves steadily from the past to the present, and on to recommendations for the future.

The Artificial Intelligence Challenge to Ethos in the Research Review

As we've already addressed in Chapters Two and Four, STEM writers (and all writers) must contend with the incessantly advertised promise of artificial intelligence (AI) tools to substitute their large-language data for your own thinking and writing. If readers suspect that your work is really the product of a bot, your credibility and your claims of originality are damaged, sometimes severely. In Chapter Two especially (pp. 63–65), we give advice about how you can protect your ethos as a researcher and writer, while also using AI to generate **first-draft material** that you can then

- fact-check and scrupulously analyze for accuracy, and
- revise extensively to fit your own purposes and data, as well as the perceived needs of your intended readers.

You may also, as we note in Chapters Two, Four, and Thirteen, use AI to proofread your final draft, but

- again you must carefully check the AI language for accuracy and for appropriately addressing your intended readers.

For a research review, you should never accept at face value what AI presents as a list of sources and surely not accept what an AI tool might say about the *value* of a source. But you could use an AI response to provide a **starter list** of possible sources that you might add to a more comprehensive list for your own scrupulous analysis and careful reading.

Further, if you are creating a review that fits the Type B format, be sure to describe in your methods and materials how you might have used AI as part of the process of discovering sources that bear on your research. Noting the ways by which you have carefully used AI to arrive at this preliminary discovery can help assure your readers that the work is thoroughly yours, and that in no way has AI done your thinking for you.

PURPOSE IN THE RESEARCH REVIEW

As described at several points in this chapter, the two basic purposes of the research review are

- to influence how readers understand the implications of this research progress and
- to influence how this research will be carried on in the future.

But as you plan to write your research review, you can add to this list of basics **more specific purposes** that might pertain to your subject and to your readers.

For example, if you know that theorizing about your line of research has featured a **specific controversy** that continues to bedevil researchers, then one additional purpose of your research review will be to address this controversy. You will perhaps wish your review to argue for one of the dominant interpretations—or perhaps to offer an alternative theory that seems to you to emerge from the research literature.

A second example would involve what you might regard as a **misinterpretation** of the research or a **popular misconception** that has sprung up among a segment of the audience that you want to address. In Chapter Four, to cite a specific instance, we looked at the discredited, but still widely held, notion that vaccines for childhood diseases cause autism. If you are addressing a less specialized audience on this topic, you may wish your research review to address this misconception directly.

In whatever ways you want your research review to achieve either the general goals of the review or the more specific purposes you identify, spend time while planning your review to write about the purposes you wish to achieve. As with all other kinds of "writing to learn" (see Chapter Two), this writing exercise will help you shape not only what you want to say, but how you will structure your argument. It will also help you choose the studies that you will want to be sure to review.

Exercise 7F: Write informally about a line of research in which you are particularly interested, and about which you might want to write a research review. For your writing to learn, try addressing these topics and questions:

- Describe the line of research you want to review. Be as specific as you can, narrowing your scope so that you won't have an impossible number of studies to review. Doing a conscientious keyword search of databases in your field should give you an idea of how broad your topic should be to be manageable.
- What purposes would you like your review to achieve (including the general goals of the research review)? Identify specific purposes, such as those suggested above, plus any others that come to mind.
- How does doing this writing-to-learn exercise help you identify your purposes?
- How has it helped you modify your focus on a line of research?

AUDIENCE IN THE RESEARCH REVIEW

The research review allows a huge range of intended readers, all the way from your fellow researchers, who are intimate with your proposed line of inquiry, to non-STEM readers barely aware of your topic. How you imagine your intended readership of the review you wish to write will help determine *both* how **broad or narrow** your review can be and the **purposes** you wish it to achieve.

As we've already seen, the research review can directly focus on an audience of fellow researchers in the narrow line of inquiry addressed. The Type B example concerning the effects of pesticide residues on bees contains the extensive methods and materials and results sections only intelligible to researchers in the specific field.

In contrast, the concise Type A example concerning DNA tomography in the cell nucleus can be useful to researchers in DNA imaging, but it can also serve as an introduction to the problem of DNA imaging in vivo for readers not previously aware of the current methods or their difficulties. Because the research review always begins with a summary of the history of research on a given question, it can usually help those with other backgrounds begin to understand a line of inquiry.

Lastly, the book, play, video, etc. (Type C) on a STEM topic can employ a wide range of styles, tones, and genres (see Section II above). Therefore, the work can be addressed to an astonishing range of intended readers. Further, the work might be sufficiently long—and usually multi-chaptered—so that different sections can be primarily addressed to different groups of readers, if the writer so chooses. Even a book designed primarily for children can be an introduction or a refresher for adults with non-STEM backgrounds.

AUDIENCE SPLITTING IN THE RESEARCH REVIEW

As described in Chapter Five (pp. 129–30), the different sections of the IMRD research article can focus on different readers, if the writer (and the journal editor) wishes. The introduction and/or the discussion sections can be addressed to broader audiences, while the methods and materials and results sections can be meant only for fellow researchers in the line of inquiry. Audience splitting is also easily possible in Types A and B of the research review. Within the template of features of the review (pp. 177–78), the introductory history of research can be easily addressed to a broader audience, as can the final sections that offer implications and recommendations for further research.

ADAPTING YOUR RESEARCH REVIEW TO DIFFERENT READERS

Definitely review Chapter One (pp. 31–35) and Chapter Five (pp. 125–30) for advice on reaching readers with different levels of technical knowledge and vocabulary, as well as with differing expectations and interests.

Those chapters also include exercises that will help you become a more discerning reader of STEM materials, as well as become able to apply your enhanced reading ability toward being a more skilled adapter of your prose to meet these different reader expectations.

GETTING GOOD FEEDBACK

Always keep in mind that writers should cultivate a few readers (usually fellow writers) who can give them feedback on their drafts. Chapters One and (especially) Two offer advice on ways to select and cultivate such readers (and how to show them respect) and how to get the most from their assistance. Such feedback can be particularly valuable if you are working in a new genre, such as the research review, in which you have little confidence.

If you are trying to reach an unfamiliar audience, it is especially important to try to **choose helpful readers with a level of knowledge similar to that of your target audience**—or a fellow writer who has experience writing for that audience. Don't choose, say, a fellow research team member just because that person is conveniently available. If you are a student lucky enough to be in a writing class with fellow students from a variety of majors, you may be able to identify classmates who might prove good helping readers and who come from the reader groups you are trying reach.

It also bears repeating that writing for readers with less technical knowledge and vocabulary than you possess should always be respectful. "Dumbing down" is never an appropriate or respectful metaphor for the **translation process you need to learn in order to reach readers less familiar to you and less knowledgeable in your field**.

Keep in mind that these readers often possess knowledge, experience, and expertise that you do not have, and that you can access in helping these readers understand and respect what you are attempting to achieve.

Moreover, always remember that the reason you are writing for these unfamiliar readers is that they can provide you with benefits that you can't receive from your fellow team members or from others in your research field. Some of these benefits include the following:

- research funding
- markets for products that might come from your research
- higher enrollments in your classes or department
- wider respect for your research field (or for science itself)
- the chance for understanding by people hostile to your research
- social and political impacts of your research
- someone other than your team members to listen to you

Exercise 7G: Continue the informal writing you began in the previous exercise. This time focus more precisely on the nature of the readers you want to reach. Answer these questions:

- Speculate on the likely interests, types of experience, and level of knowledge of your field of your intended readers. How might you become more aware of these characteristics in your intended readership?
- How can you describe your field and line of inquiry in order to interest these readers?
- What are you hoping to achieve by reaching these readers?
- Will you do audience splitting in this piece? Where? How?
- Does writing about audience in this planning document help you modify the purposes of your research review?
- Do you foresee a particular difficulty in "translating" your research for a less knowledgeable audience? How might you achieve clearer communication regarding your research and its importance?

CONCLUSION

This chapter has explored some of the versions of the research review, an underappreciated but classic and essential genre of STEM communication. We have looked at three important types of research review and offered a range of sample reviews and advice for writers.

The many options in styles, genres, and audiences open to the authors of research reviews prepare the way for Chapter Eight, "STEM Journalism—Writing, Reading, and Connecting with Broader Audiences." This chapter will challenge STEM students and professionals to practice and learn writing and design skills that can open them to exciting possibilities for their communication of research—and their advocacy of science.

Using the multi-modal ideas introduced in Chapter Three, the next chapter will help readers of this text make the transition from solid writers of traditional STEM genres to researchers who develop their creativity to communicate in new ways to new—and many more—readers.

CHAPTER EIGHT

STEM JOURNALISM—WRITING, READING, AND CONNECTING WITH BROADER AUDIENCES

Topics in this chapter:

I. Thinking of Yourself as a "STEM Journalist"
II. Who Are Your Readers and Why Do They Care?
III. Writing Your STEM Popular Article—Tips on Voice (Ethos) and Organization
IV. Tips on Style, Vocabulary, and Ethical Considerations

Popular writing about all STEM fields has exploded in the twenty-first century, the result in part of online, multi-modal technology. More than ever, STEM students and professionals have opportunities to reach broad audiences—through print media such as newspapers and magazines, but more easily via online versions of those same publications, plus lab websites, blogs, and social network communities. Indeed, these opportunities are so numerous and so accessible that most STEM writers are feeling the pressure to get out there and write about their research for audiences they never thought they'd have to reach.

What this means is that most STEM writers have to get out of their comfort zones. They need to drop the acronyms, technical jargon, and strings of math formulas. They need to learn the wants and needs of very different readers from those they communicate with in the lab, the field, the IMRD journals, and the conference halls. They have to become familiar with language—vocabulary, styles, metaphors—that they may have scoffed at as "not science," and perhaps "dumbed down" for a mythical "general public."

We've been through this need for change elsewhere in this book, especially in

- the Introduction and Chapters One and Three, where we explored multimodal tools that are increasingly necessary for STEM communication;
- Chapter Four, where we probed the tension between the rhetorical desires of non-STEM readers and the scientist's obligation to honor the often-tedious thoroughness of the scientific process; and
- Chapters Five, Six, and Seven, where we delved into writing for peer-reviewed journals and saw how those venerable publications are ever more aware of other readers. We also saw how STEM researchers often have to split their rhetorical attention between their fellow researchers and those many, many others whose interest in STEM research comes from other reasons: personal, political, ethical, economic.

But here we look closely at those STEM-relevant publications that can reach huge readerships, often in the many millions online: publications such as *National Geographic*, *Canadian Geographic*, *Scientific American*, the *New York Times*, *CNN*, *CBC*, and other large news sources. From looking closely at strategies used in these publications, we can learn techniques that can be applied in smaller, but still important, STEM publications, such as professional and political **newsletters**, research lab **websites**, and **brochures** that strive to bring scientific research to

- voters,
- legislators,
- professionals in business and industry, and
- consumers of all types.

The key question of this chapter is "**How can STEM students and professionals contribute to this vital purpose of STEM communication?**" As always, each section will look at sample writings and present exercises to get you into the game.

I. THINKING OF YOURSELF AS A "STEM JOURNALIST"

You might consider this a strange and unappealing idea. "Me? A journalist? Never!" If you are like many of the students we've had, you see your comfort zone as the lab or fieldwork site you work in (or want to work in), and you think of your writing as consisting of notes and calculations, reports to your boss, lots of emails and texts among colleagues—and maybe a contribution to an article for an IMRD journal in your field. Those tasks are tough enough.

When you think of "other readers," you might think first of the friends and relatives who claim to be interested in what you do, but whose eyes glaze over quickly when you try to explain it. The more excited you get when you go into the details of what you do, the more bored they look. You surely don't envy your boss who has to try to convey a sense of what your team does to the people who may be funding your lab, but who just want to see results, and who don't want to hear "Well, it's more complicated than that."

STEM geek that you are, you pick up from time to time popular STEM-relevant magazines, or click on catchy titles in news source websites about the latest new discoveries in medicine, astrophysics, or paleontology. But when you read the articles, you always feel that so much has been left out. "It's more complicated than that!" you scream to yourself. Indeed, you feel that in the quest to make a big splash to "the public," all the interesting stuff has been ignored: the quest for accurate methods and the right materials, the surprising cases or artifacts that just don't fit what you expected, the conflict of interpretations among researchers.

Even worse, you fear that "the public," whoever they are, are being taught that science is about the next breakthrough, the next wonder drug, the next whatever that will solve the problems of the world and will soon be streaming live on your phone. It's never about the tedious process, the false starts, the time—and money!—it takes to do things right, the questions that just get tougher to understand.

THE JOY AND CHALLENGE OF WRITING FOR A DIFFERENT AUDIENCE

If this scenario is somewhat like yours, then why, we wonder, do almost all our students get most excited when we ask them to try to "translate"—that's the word often used—something that thrills them in their scientific work for a non-specialist audience that they identify as people they want to reach?

When our students write their proposals for this "Popular Science Project," we ask them to write about their ideas **rhetorically**:

- to describe their topics in terms of the **purposes** that they'd like their **writing and design** to achieve,
- to describe as closely as they can the **types of readers** that they'd like to reach,
- to describe the **genres** and the **media** they'd like to use.

Some of them imagine writing feature articles for mass circulation magazines or news sources—*National Geographic*, *Time*, *Elle*, *The Wall Street Journal*, the *New York Times*. Others design six-page brochures for the display racks in stores, medical offices, or tourism sites. Still others build several-page websites (with videos), or create multi-slide PowerPoints to accompany their addresses to school audiences, or write and draw children's books, or design posters for student academic exhibitions.

Their research topics are as varied as the many STEM majors on our campuses. In every case, the assignment excites them because they see it as a rare opportunity to really think about how to

- pique the interest of non-STEM people whom the student feels can use—and need—the information the writer will share, and
- experiment with language, style, genre, and media—photography, video, drawing, infographics, etc.—that they have read and seen, but which they themselves have not had the chance to apply to their STEM research interests.

They prepare for these projects by studying examples of the genres in which they want to work, much as we do in each chapter of this book, and in the exercises each chapter presents. They also consider the interests of the audiences they want to reach, and strive to adapt language and evidence that their intended readers will respond to. In Section II of this chapter, we will study these rhetorical strategies in STEM journalism.

THE CREATIVE CHALLENGE FOR STEM PROFESSIONALS

Many STEM journalists are trained writers with no professional STEM background, but who collaborate with STEM professionals and learn from them using the strategies of journalists to perfect their craft. Some of these strategies include

- interviews;
- quotation and dialogue;
- first-person story, reflection, and analysis;
- vivid description of places and processes;
- visual (and audio/visual) media; and the
- the study of audiences.

But these techniques can be learned by STEM professionals, as well as by STEM students. The many STEM-trained writers who have written for popular publications have learned these techniques through practice. This chapter is meant for all writers who want to reach these broad audiences. Like our STEM students, more and more STEM professionals have seen not only the need to reach diverse non-specialist audiences, but also enjoy the creative challenge of working in new genres and using new

- language tools,
- media, and
- types of evidence.

In Sections II, III, and IV of this chapter, we will study samples of STEM journalism, in which we will focus on language tools, media choices, and types of evidence that writers of whatever educational background use to reach broader readerships.

II. WHO ARE YOUR READERS AND WHY DO THEY CARE?

To answer these essential questions, we explore the first sample article: the lead story of the June 2024 print issue of *National Geographic*, titled "The New Science of Stress" (Bhattachargee 2024):

> The defense mechanism affects our health at every age—and researchers are striving to understand exactly how. (17)

This highly photographic article covers 34 pages, of which just 3 are text only. Almost every other page contains some text, but all are dominated by captioned, multi-color photos of people (and a few animals) in a wide range of situations and work scenes: for example,

- a black lab mouse searching for escape from a maze
- armed, helmeted law enforcement agents in an active shooter training session
- a family in Georgia (US) with three small children in a small, crowded kitchen
- gloved hands dissecting a slice of mouse brain tissue
- a young woman in a hijab about to go through an MRI machine at the Harvard Stress and Development Lab
- a New York City public health worker giving advice to a new mother "about breastfeeding and good sleep habits for newborns" (30)
- a small bird being held by a "Tufts University doctoral student" (32) who is about to take a blood sample

- an elderly woman and her bed-ridden, Alzheimer's-afflicted husband in "an assisted living facility in Ohio" (34)

plus a two-page **infographic** drawing of a man, his internal organs delineated and his right hand holding his bowed head. The infographic shows numerous bulleted sentences describing aspects of the "Chronic Stress Effect" (28). In addition, page 3 of the fold-out that begins the article is a widely used "Stress Level Quiz" (17), which the reader is invited to take.

Who are the intended audiences for this article? Clearly there are many. The first-person "our" of the subtitle is meant to include any and all potential subjects, including all the diverse kinds of people and animals pictured and described in the photos. The intent is to say, "We're all in this crisis together," as "we" seek to deal with the many sources of stress in the world today. Indeed, even before the text of the article begins, large-font, bolded words on a red background (16–17) work as a kind of abstract of the article:

> Over the past few years, the perceived level of psychological stress has risen dramatically across the globe due to a combination of events, including the long-lasting COVID-19 pandemic, civil unrest, escalation of political instability across the globe, and climate change that has triggered major environmental and economic perturbations. The consequences of this broad-scale increase in stress are only beginning to be appreciated. But evidence suggests that we are facing a second pandemic of mood and anxiety disorders, including major depression, anxiety, and posttraumatic stress disorder. (Akil and Nestler)

This abstract, quoted from a prestigious journal, is immediately followed by the "Stress Level Quiz," and so the reader is invited, almost pulled, into the article. Turn the page and you are hit with the title of the article itself in huge-font white and red letters on a black background: **THE NEW SCIENCE OF STRESS**. So design and wording combine to create a scary mood intended to be stressful. Is this journalistic tone-building unfair to the reader? Unfair or not, it certainly contrasts with the much more subtle tone-manipulation of the typical IMRD journal article.

Detecting tone. One good way to detect the attitude that writers take toward an audience is by studying **tone**. Recall that we explored tone in writing in Chapter One (pp. 41–42) as part of style and tone, one of the six categories of rhetorical analysis. Tone can change within any piece of writing, and when we detect a particular tone toward an audience, we do so by looking at **what is said** about a group and **how it is said** both in words and in graphics.

Major themes developed in the article. Following this abstract, the quiz, and the colorful title-page, the multi-modal article goes on to elaborate on these three themes:

- Everyone, of any age and background, suffers from multiple stress factors.
- Researchers in multiple fields are seeking to understand stress through careful testing.
- Health professionals in diverse disciplines are attempting to help "us" combat stress.

Two of the major ways that the writer, Yudhijit Bhattachargee, uses to develop these three themes is by (1) **describing many studies** and (2) **quoting from interviews** with a series of neuroscientists, endocrinologists, etc., who work in diverse forms of stress, from workplace pressures, response to loud noises, and environmental disasters, to deaths in a family, break-ups of relationships, and a whole host of common fears. So many are the forms of stress mentioned that it would be almost impossible for any person not to be subject to at least one of them.

The **photography** and **formatting features** of the article contribute strongly to these emphases. The formatting feature known as the **photographic caption** is prominently used to focus reader attention and attract readers. Indeed, the captions not only give information about the people and places in the photos, but also each caption has a bolded title in red that heightens reader concern. For example:

- "Memory Loss,"
- "Taking a Toll,"
- "Invisible Impacts," and
- "Changes to the Brain" (Bhattachargee 2024, 18–47)

The large size and interior settings of most of the photos (by photographer Brian Finke) also pull readers into the scenes.

These photos and others show stress as a highly varied phenomenon. In several of the photos, we see clinical settings, where professionals work with patients. These photos and captions reinforce the link in the article between the increasing presence of stressors and the clinical effort to find workable treatments.

A multitude of intended readers. The variety of international examples described in the article, plus the variety of forms of stress, illustrate the worldwide spread of these phenomena. On the other hand, that all the photos show settings in the US—primarily in the Northeast—means that readers who look primarily at the US-based photos and captions will get a somewhat different impression of stress from that of readers who actually delve into the much more international text. Keep in mind that these choices by writers/designers help determine the **potential audiences of interested readers** for this article. If you consider the age range, the settings described, the causes of stress, and the demographics affected, the list of groups of potential interested readers is huge. Does a writer attempt to reach them all?

When we and our students analyze articles like these for intended readers, we note such features. We consider why this breadth—or lack of breadth—of readers would be interested in these articles, and why and how the publication and its authors attempt to appeal to all or some of those groups of readers.

Attraction of readers vs. respect for readers. When we begin listing the groups of readers, we study the article to see what it contains that might not only attract these readers, but also how the article might raise questions for them or even opposition.

How do writers treat their readers? Asking such questions about readers pushes us to look more closely at how the writers treat their readers in the text and in other media they use. How they write about different types of people can show if they have respect for these readers, or if a group of readers comes across in an article as a **stereotype** or an **object of criticism** or even contempt.

For example, in the article on stress, we've noted that one of its major themes is that stress is unavoidable, inevitable, and everywhere. Individuals cannot avoid being stressed, and everyone is respected for having to deal with this common plight. Further, the writer very much respects the many scientists and clinicians who strive to understand the causes and effects of stress, and who are seeking to find treatments to ease its impacts.

On the other hand, how individuals and groups of people, including whole societies, choose to handle stress in their lives is either respected or criticized in the article. For example, the list of "stress busters" in the infographic recommends specific lifestyle choices:

- Supportive relationships
- Balanced nutrition
- Physical activity
- Time in nature
- Regular sleep schedule

- Relaxation techniques
- Mental health care (Bhattachargee 2024, 29)

Each of these choices will be much easier for individuals with enough financial freedom, time, and family stability to make them. Does this mean that readers who look at such choices and find them extraordinarily difficult to achieve are being silently excluded from the intended readers? Are they being silently told that healthful ways of dealing with stress are unavailable for them?

The last eight pages of the article focus on one class of stress reduction methods: mindfulness meditation. Indeed, the examples described do show that the breathing and visualization techniques have significant benefits in such elements of stress reduction as pain tolerance, effective brain function, and relief from anxiety. Once again, however, the techniques are described as a matter of individual choice. There is no mention in the article of government or privately funded action to make such choices open to people who cannot afford them. Again, respect is reserved for those who make that choice. Those who cannot or do not choose these healthy options are not respected.

So, from a rhetorical perspective, readers are being challenged to become part of the accepted group or remain outside.

Exercise 8A: As a first exercise for this chapter, identify a topic about which you might like to write for an audience of non-specialists, particularly those without STEM backgrounds. Write informally for yourself about this topic, so that you can identify a message that you'd like this audience to understand about this topic. What would be the purpose of your writing to them? How would you hope to influence their thinking or behavior in relation to this topic?

Write also about the audience: what are its characteristics? Why do you think they might be interested in your message? Why do you feel that they should be aware of your topic and message?

Now consider a genre or type of writing that you think would be appropriate and effective for conveying this message and achieving your purpose for this audience. What media (e.g., photos, tables/charts, video, drawings, etc.) would you want to include to help you achieve your purpose for this audience?

Exercise 8B: Choose an article of STEM journalism from either a large-circulation magazine (print or online) or from a large-circulation newspaper (print or online). (Recall, for example, the article on endangered species from the *New York Times* that we analyzed in Chapter One.)

Perform an analysis of potential audiences for the article you choose. Use the methods for identifying audiences that we've explored in this chapter to this point. Based on what you read in the article, make a list of possible audiences. If your list is long, which it might be, focus on two of these potential audiences. How is tone used in the article to show the attitude of the writer(s) to each audience?

Study the article also to discover why each audience might want to read it: why should that audience care about this article? What have the writers done to help that audience care about and perhaps benefit from the article?

In the article you choose, do its writers/designers show respect for the audiences you've identified or seem subtly to be criticizing them?

If you were revising this article, what might you do to increase the likelihood that that audience would care about the article?

III. WRITING YOUR STEM POPULAR ARTICLE—TIPS ON VOICE (ETHOS) AND ORGANIZATION

If you have not yet completed Exercise 8A (above), do so now, as it will be important in your reading this next section on **voice (ethos)** and **organization.** Follow-up exercises in this section will help you practice the ideas and techniques described here.

VOICE AND ETHOS IN YOUR STEM POPULAR ARTICLE

In several chapters thus far, we have considered the vital role of ethos in STEM writing. By this time, you know that ethos is the technical rhetorical term for what we as readers think of as the **character, reputation**, and, to some extent, **personality** of the writer. In most cases in STEM journalism, the name of the writer is not particularly well known, unless the writer is also a renowned scientist. In those minority cases, the character and reputation of the writer is known before readers encounter the article, and so we as readers don't depend on the article itself to give us a reason to respect or admire the writer—or to think of the writer as a controversial figure.

No, in most cases of STEM journalism we may even disregard the name of the writer. We plunge into the article and judge the quality and argument of the article through our reading. In contrast, in the stress article from *National Geographic*, we are told at the outset that author Yudhijit Bhattachargee has written many features for the magazine and is a book author on a related topic. Late in the article itself, Bhattachargee relates a brief story about growing up in India and making visits to an older relative, who was a daily meditator. Photographer Brian Finke is described in small print at the end of the text as having worked on three other articles for *National Geographic*. So although we do get some early sense of the quality of what we might read in the article, it is only by reading the article itself that we can properly judge it.

When we discussed ethos in Chapters Five through Seven, we made the very similar point that our judgment of the quality and value of the research article and the research review depend on how well the writers **demonstrate their knowledge** of the issues and of the research and on how well they **argue toward their conclusions**. Here, too, in STEM journalism we as readers will judge the work on the evidence that is brought forward and on how well the writers argue for their points of view.

However, as we shall see in this section, the types of evidence used in STEM journalism are quite different from those used in IMRD experimental articles, as are the typical parts of a journalistic article and their organization. We'll explain these differences below.

WHAT IS "VOICE"?

Also important in STEM journalism is what we call the "voice" of the writer. Voice is an old term in the interpretation of reading, but it's still somewhat ambiguous. Why? Because most of our reading is silent, yet we often imagine someone speaking in our heads as we read, as if there were not only an actual voice, but also—and this is the important point—a "person" speaking with some personality and trying to communicate a particular tone.

Why Is Voice Important?

Voice is vital in STEM journalism because, whether we are aware of it or not as we read, writing that has personality is much more likely to keep you reading than writing that comes across as "flat." But since we are reading silently, all that personality has to come through in the words on the page or screen, and through such things as punctuation and sentence length. (More on these ideas below and in Chapter Twelve).

Finding Voice in a Sample Article

Let's look at this article (Broad 2017) from the online *New York Times* to see how voice works in a piece of STEM journalism.

A Giant Nuclear Blast, but a Hydrogen Bomb? Too Soon to Say

Not all nuclear arms are created equal. They have four main designs that progressively raise the destructive power of the weapons and their ability to obliterate large targets. The biggest are known as city busters.

Despite that range of fiery outcomes, and the unthinkable consequences, nuclear experts say that unusually large test explosions can be achieved in a variety of ways, making the field of atomic forensics quite difficult for distant experts.

With Sunday's large North Korean blast, Norsar, short for the Norwegian Seismic Array, a research group based in Kjeller, noted that global shock waves alone provide insufficient information to determine if the detonation was a true hydrogen bomb, as the North declared.

"But we can say in general," the group added, "that the credibility of the claim increases with increasing explosive yield." It said other evidence, such as leaks of radioactivity from the underground test site, might solve the riddle of the bomb type in the weeks to come.

Nuclear experts said the blast on Sunday was somewhere between four and 16 times as powerful as North Korea's previous largest explosion, which was about the size of the Hiroshima bomb.

The most basic kind of nuclear weapon—the kind experts say the North began with—is known as an atomic bomb. It gets its energy

North Korea's Program Is Probably at an Intermediate Phase of Development

The secret to achieving more destructive power is to increase the amount of thermonuclear fuel that an exploding atomic bomb can ignite. North Korea said that Sunday's test was a hydrogen bomb, but analysts were skeptical of this claim.

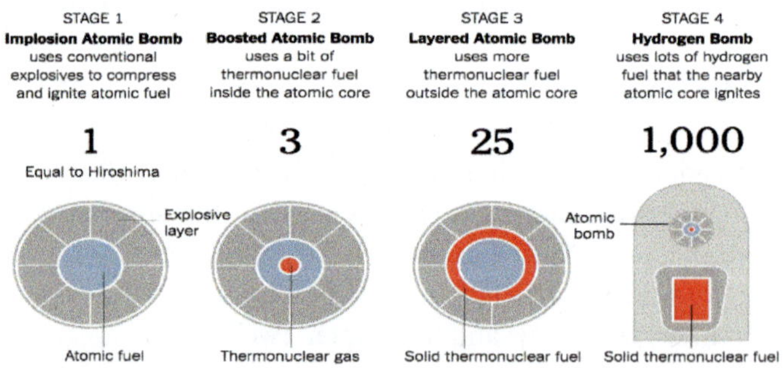

Note: Destructive power for each stage is based on early tests in the U.S. and U.S.S.R., not on current stockpiles.

By The New York Times

from splitting heavy atoms in chain reactions. The main fuels of atomic bombs are uranium or plutonium, both heavier than lead and both far more expensive than gold.

A cheaper and more powerful way of freeing the atom's hidden energies is to fuse two light atoms into one. The main fuels are deuterium and tritium, both rare but inexpensive forms of hydrogen. They are known as thermonuclear fuels because their ignition requires the blistering heat of an exploding atomic bomb, which acts like a match.

The United States in 1951 injected a tiny amount of thermonuclear fuel into the core of an atomic bomb, enhancing its power. The explosion was roughly three times as strong as the Hiroshima blast.

Some experts say the North is now taking this second step in bomb design. Others say the isolated state may have raced ahead to the third stage. That method wraps alternating layers of thermonuclear fuel and uranium around atomic bombs, and burns more hydrogen than simple boosting.

David Albright, a former United Nations weapons inspector and the president of the Institute for Science and International Security, a private group in Washington that tracks nuclear arms, noted in a recent report that Russians who first tried that approach produced a blast more than 25 times as strong as the Hiroshima bomb.

A true hydrogen bomb, the fourth stage of development, works by positioning near the triggering atomic bomb a separate capsule that can hold a much larger amount of thermonuclear fuel.

In 1954, on Bikini Atoll in the Pacific, the United States tried that approach. The fireball expanded for miles. The shock wave swept neighboring atolls clean of vegetation and animals. In minutes, the mushroom cloud rose some 25 miles. Slowly, its radioactivity spread around the globe.

The destructive force of that single hydrogen device turned out to be far greater than all explosives used in World War II, including the atomic bombs dropped on Hiroshima and Nagasaki. The blast, codenamed Bravo, was 1,000 times as powerful as the Hiroshima bomb. It was the nation's most violent thermonuclear test ever.

Experts say North Korea is unlikely to achieve that kind of destructive power anytime soon, if ever. But they say the accumulating evidence suggests that the reclusive state is trying to do so—and trying hard.

First, note the link to the **author's bio**, which identifies Broad as a senior science writer for the *Times* and a two-time Pulitzer Prize winner. These credentials establish the writer's ethos for those readers who care to look at the bio. But most readers will ignore this link, in part because it requires the reader to click on the author's name in very small print. Most readers will just jump into the timely article (in 2017) on North Korean nuclear testing. So how does the article establish ethos and create a voice for the writer?

The title itself builds toward voice in the article, because it starts off with a question: "... but a hydrogen bomb?" followed immediately by a response: "too soon to say." The **structure therefore imitates a conversation**, and the reader unconsciously imagines two people talking. The use of what is called a "rhetorical question" is one of the oldest and easiest ways to build voice in prose writing.

Then look at the first two very short paragraphs:

> Not all nuclear arms are created equal. They have four main designs that progressively raise the destructive power of the weapons and their ability to obliterate large targets. The biggest are known as city busters.
>
> Despite that range of fiery outcomes, and the unthinkable consequences, nuclear experts say that unusually large test explosions can be achieved in a variety of ways, making the field of atomic forensics quite difficult for distant experts. (Broad 2017)

The very first sentence is very short, and is written with **no technical jargon**. The writer is aiming for a broad readership. Many US readers will see that the sentence is ironic, with an echo of the US Declaration of Independence ("All men are created equal"). The reader imagines a speaker who is serious, but at the same time a bit playful with language.

The next sentence shifts voice and length—this is a more **technical, quantitative voice**, with larger words. But then sentence three brings us back to the first voice—and ends with a bang: "The biggest are known as city busters" (Broad 2017).

Lastly, the second paragraph gives us the formal, academic voice again, and is just one long sentence. More importantly, it brings in the first piece of **evidence**: a reference to what "nuclear experts say" about "unusually large test explosions." The sentence ends with, not a bang, but what works as a **thesis statement** for the entire article: "making the field of atomic forensics quite difficult for distant experts" (Broad 2017).

So, just by (1) varying sentence length and (2) the limited use of technical vocabulary, plus (3) the rhetorical question and (4) a bit of irony, Broad has created voice and personality in the article—in just a few words in two short paragraphs.

One more simple way to add voice and personality, which Broad uses, is through **direct quotation** and, less effectively, through **paraphrase**. In paragraph three, Broad introduces his main source, Norsar (Norwegian Seismic Array), a research group whose report on the blasts in North Korea is linked to the article. In paragraph four, he quotes from the report. Although this quotation is technical, just the use of quotation marks gives the reader a sense of the spoken word. Perhaps more effective at creating voice is the repeated use of the words "say," "says," or "said" before a paraphrase: e.g., "North Korea said that Sunday's test was a hydrogen bomb...." These "says" words are used nine times in the brief article, including in the title.

EVIDENCE AND ORGANIZATION IN STEM JOURNALISM

Early in this chapter (p. 200), we looked at typical strategies used by STEM journalists that **differ from the conventions** of the IMRD article. These tools included

- interviews,
- quotation and dialogue,
- first-person story, reflection, and analysis, and
- vivid description of places and processes.

These strategies actually serve as **types of evidence** as writers develop their ideas and argue for their main points of view. Along with these strategies, writers build their cases using three other types of evidence that are also used by IMRD writers:

- statistics,
- visual (and audio) media, and
- (in online publications) links to sources.

The lead article on stress from *National Geographic* that we studied earlier gives examples of all of these tools. Bhattachargee and photographer Finke collaborate to create a **highly visual argument that uses many voices**—through quotation, personal narratives by those they've interviewed, and verbal descriptions that reinforce the impact of the photographs. For example, the following paragraph uses description of process, quotation from an interview, and personal story:

> Jorge had heard about meditation years earlier while in a wrestling club. Somebody he knew there would meditate before matches and didn't seem to get agitated, even when losing. "I used to wonder, how

come I'm always mad and he's always calm?" Jorge said. The question had been partly answered, he said, by what he had been learning in the prison meditation program. And he began applying that to his life. "When I feel like I'm getting frustrated or angry, I just breathe," he told me. (Bhattachargee 2024, 47)

In contrast, Broad's brief article from the *New York Times* on the North Korean nuclear testing concentrates its argument on two forms of evidence:

- **links to sources** from the Norwegian Seismic Array research group and other scientists, which the writer **paraphrases**, and
- a relatively simple **infographic** that portrays through schematic drawings the four major "stages" of development from the atomic bomb to the hydrogen bomb.

Exercise 8C: Return to the article you chose for Exercise 8B. This time, concentrate on two factors: voice and types of evidence.

Using the above description of ways to create voice and personality in STEM journalism, identify how these qualities are created in your chosen article. What techniques stand out for you? Jot down examples from the article.

Similarly, using the above description of types of evidence in STEM journalism, identify the tools and technique used in the chosen article to build evidence. Which tools are used?

In reflecting on your analysis, consider if there are techniques that the writer uses that are not covered in the above descriptions. If so, how would you characterize and name those tools?

ORGANIZING YOUR STEM JOURNALISM ARTICLE

Unlike the IMRD journal article, STEM journalism offers a wide array of organizational possibilities. Certainly, within STEM journalism there is nothing like the strict IMRD format of abstract, introduction, methods and materials, results, and discussion. In Chapters Five, Six, and Seven, we encountered a few variations on this organizational pattern, some of them brought about by advances in information technology, but any STEM-trained writer knows to look in IMRD journals for the basic structure and the kinds of information in them.

But STEM journalism doesn't play by these formatting and organizational rules. In just the two pieces we studied so far in this chapter, we saw the many-paged *National Geographic* article organized by

- a series of captioned photographic displays of different settings, each depicting a different kind of stress, research, or treatment, and
- its pages of text similarly organized by a series of personal stories and research descriptions and by a progressive focus on how stressors affect people of different age groups (childhood, adolescence, adult, elderly) and the means by which researchers/practitioners try to reduce the effects of chronic stress.

We then studied briefly the very short article from the *New York Times* organized to answer a controversial question about a nuclear device exploded in North Korea. This article is organized as an argument dependent on research from two paraphrased sources, but with

- no personal stories, and
- only one graphic device, a simple diagram of four stages of nuclear weapons development.

As you study varied examples of STEM journalism, you will encounter many ways that writers and publications organize and format their work, as they use the many tools of the craft, including those that we listed earlier in this chapter (pp. 211–12).

Nevertheless, what all STEM journalism has in common is a **motivation** that it shares to some degree with IMRD publications, but that is much less obvious in STEM journals: **the need to attract intended readers in a highly competitive environment**.

Because STEM journalism does not have a highly conventional format, as do IMRD journals, competition in STEM journalism has led to much experimentation with formats and organizational patterns, all in the service of drawing readers (and viewers) to pay attention. Moreover, the lack of a conventional structure and the need to attract greatly diverse readerships has given rise to the great variety of media tools used in STEM journalism. Popular STEM publications have set the example in science of having both print and online versions, as well as multi-media websites and even (in the case of *National Geographic*) television networks.

Exercise 8D: Go back to the article you chose for Exercises 8B and 8C. This time look for its organizational pattern. How would you classify its opening paragraphs? For example, does it begin with some type of abstract, one that perhaps looks forward to a conclusion (as the *New York Times* story does)? Does it begin with a personal story that in later paragraphs the writer turns into a generalization? Or does the writer choose some other form? If so, how would you name that organizational pattern?

Proceed through the article and try to give names to its different sections. List them in an outline or flow chart. This will help you see the article's overall schematic design.

Bottom line: Does this structure work for the article in terms of attracting you as a reader and holding your attention? Or do you feel the need for something different that can make the article more powerful or clearer? What might that "something different" be?

TWO COMMON STRUCTURES IN STEM JOURNALISM

Nevertheless, keep in mind that STEM journalism, just like the IMRD article, is governed by the need to **argue convincingly based on evidence**, and this means that much STEM journalism will carry out arguments in ways that readers often will find familiar. Two of these common structures are

- the personal story, and
- the explicit argument.

The Personal Story

> Khorakiwala describes herself as a worrier. When I spoke to her last September, she was feeling stressed because she'd missed a couple of days in the week prior. "I had a lot of work to catch up on, plus after-school activities. It got too much. I felt like I didn't have any time." (Bhattachargee 2024, 30)

This paragraph comes from the Adolescent section of the stress article. The personal story is an age-old tactic of writers across many genres. Readers tend to focus intently on other people in the midst of critical situations. By highlighting the person's emotional state, the writer can then move from the narrative to a larger point that can then become an argument. In this case, Bhattachargee moves immediately from the student's expression of anxiety to a description of a 150-person study led by a University of Oregon psychologist. The study aims to

> measure how common everyday stressors that teenagers experience affect their emotional, cognitive, social, and brain development. (30)

As you read other examples of STEM journalism, keep track of how often an article begins with a personal story—which of course need not only involve humans. Note how this article from *Scientific American* (Little 2017) begins by taking us far beneath the ocean in 1987 to witness, with the crew of the submersible *Alvin*, a sight few humans had ever seen:

> The creature appeared to have been dead for years, but the bones and their surroundings teemed with life—wriggling worms, centimeter-size clams, little snails and limpets, and patches of white microbial mats. (14)

Besides "zooming in" as if in a video, note how the prose almost "moves" with the use of the adjective "wriggling" and the specific names of other small creatures that "teem" in the bones of the carcass. Again, the personal story captures our attention and makes us curious to learn what this strange sight means. We are in the writer's grip.

The Explicit Argument as Organization in STEM Journalism

> Despite that range of fiery outcomes, and the unthinkable consequences, nuclear experts say that *unusually large test explosions can be achieved in a variety of ways, making the field of atomic forensics quite difficult for distant experts.* (Broad 2017; emphasis added)

STEM students and professionals will recognize the italicized statement as similar to what they will find in an IMRD introduction, in that they will expect the article to proceed to an explanation and justification of that statement. Indeed, in the STEM journalism article on North Korean nuclear testing, that is what they will find. And that argumentative structure is typical in much STEM journalism.

In the *New York Times* article, writer William Broad proceeds from the statement above to elaborate on *each* of the two parts of the statement:

- First, he provides **details** and a **source** for the difficulty of determining the type of nuclear device. He cites and quotes the Norsar research team.
- Second, he uses the **infographic** of the four types of nuclear device—from the atomic bomb to the hydrogen bomb—to show both the schematic design of each weapon and the escalation of power of each (see Chapter Ten for more on creating infographics).

- Third, he provides **examples** of the use of each type of weapon from the history of nuclear testing, with a particularly vivid description of the effects of the first US tests of hydrogen bombs on Pacific islands: "The blast, code-named Bravo, was 1,000 times as powerful as the Hiroshima bomb. It was the nation's most violent thermonuclear test ever" (Broad 2017).
- Fourth, the last paragraph of the short article brings readers back to the **question** of the type of device exploded by North Korea, and concludes with a warning: "Experts say North Korea is unlikely to achieve that kind of destructive power anytime soon, if ever. But they say the accumulating evidence suggests that the reclusive state is trying to do so—and trying hard" (Broad 2017).

Exercise 8E: Find two samples of STEM journalism. One of these should use the personal story as a major technique in grabbing reader attention and holding it. The other should be clearly an explicit argument similar in structure to the North Korea testing article.

(**Note**: You may find that some articles you analyze will be hybrids of the two types: using both personal story or stories and an argumentative organization. You may certainly use such an article for this exercise, if necessary. Be sure to study how the personal story affects the way the argument develops.)

In studying the personal stories you find, how does the writer use tools such as quotation and physical description to create a vivid sense of the scene and the persons or other characters? Do you have suggestions for improving these features?

In studying the argumentative structure, how does the writer explain and justify the major points that are being argued? What types of evidence are used? Are sources quoted, paraphrased, and/or linked to the article? What suggestions might you make for improving the argument?

IV. TIPS ON STYLE, VOCABULARY, AND ETHICAL CONSIDERATIONS

Again, keep in mind that the most important **goal** of the STEM journalist is to **capture readers' attention and keep it**. We've seen how the personal story is a favored strategy to achieve this goal. A word we've mentioned several times in the chapter is "vivid": vividness implies excitement and sensory depth, a striving for richness of the reading experience. Such media as photography and video have been popular in STEM publications and telecasts for decades because of the need for vividness, for life-like, colorful, and multi-sensory experience of phenomena.

So in STEM journalism, whether in articles or on websites, writers/designers strive for a vocabulary that achieves vividness. In this section, we'll suggest ways to make your journalistic prose vivid.

But there's a catch...

We've also seen that a **careful, evidence-based argument** is another stylistic staple of the genre. It is important to understand that this is **STEM** journalism—it is not advertising, and it is not tabloid crime reporting or scandal-digging or misinformation. The readers who go to STEM-relevant magazines and the science sections of major newspapers are not looking only for sensation; they are also looking for well-crafted arguments that present strong evidence.

So your goals as a STEM writer for a broad, multi-layered audience are to

- write vividly, and
- write convincingly with good evidence.

In this section, we'll give suggestions for choosing words and sentences that help you to reach both goals.

WRITING VIVIDLY WITH SENSORY LANGUAGE

Compare the following two descriptions:

1. The creature appeared to have been dead for years, but the bones and their surroundings teemed with life—wriggling worms, centimeter-size clams, little snails and limpets, and patches of white microbial mats (Little 2017, 14).
2. The skeleton was observed to be in an advanced state of decay, as indicated by disintegration of those bones that could be ascertained from a distance of ten meters (+/- 1), perhaps due in part to their gradually having been consumed by microbes and crustaceans; moreover, many organisms, including what appeared to be larvae, other invertebrates, and mollusks (about one cm. in diameter), as well as dense agglomerations of microbes of undiscernible species, covered and in some cases surrounded the skeletal groupings on that portion of exposed seafloor which was visible to the unaided eye through the unclear water.

The first description (which we read earlier on p. 215) is more vivid for several reasons:

- The sentence is much **shorter**.
- The names of animals are more **common and specific**.
- The specific, common terms are **easier to see** in your mind.
- Each term has a characteristic that makes it **even easier to see** (wriggling, centimeter-sized, little, patches, white, mats).
- The sentence is shorter and more direct because it uses the **active voice**: "appeared" and "teemed" vs. "was observed to be" and "having been consumed by."
- Most importantly: the sentence **makes its point in a few short words**: "teemed with life."

Scientists used to writing in a conventionally "scientific" style might object to the first sample because it lacks conservative, tentative language. How could the viewers in the submersible be sure that they were seeing tiny worms, snails, clams, and limpets from a distance through the water? Wouldn't it be better to use more generic terms and note the difficulty of being sure from a given distance?

But the point of the *Scientific American* statement is that this dead "creature... teemed with life" (Little 2017, 14). That idea gets lost in the dense verbiage of sample two. **To attract and keep a wide range of readers, the main points of statements must be evident.** And in the context of this article from *Scientific American*, about the ecosystems created by dead whales who fall to the sea floor, making that point is vital.

Also, the terms of sample one are easier to see in the mind's eye—and that ease of seeing is critical to a vivid style. **Vivid language is sensory language.** You can much more easily "see" clams, worms, snails, and white microbial mats than "larvae, invertebrates, and agglomerations of microbes." If you can see those creatures, then you can more easily imagine yourself as part of that scene. The writer will engage you and keep you paying attention, provided there is that sensory connection through words.

Using All the Senses

But sensory language is more than words that we can "see." Our language in STEM journalism needs to appeal to all our senses: sight, sound, texture/touch, motion, smell, taste. If you think about it, writing accurately about sense perceptions is indeed scientific writing, because it is much more precise than the vague, abstract language that we usually substitute for the sensory.

Try the following exercise:

Exercise 8F: Choose a scene or an object in a scene that is familiar to you (from your home perhaps or a place you often frequent). Describe it in terms of your sense perceptions only—do *not* judge it or give an emotional reaction. For example, if you describe a favorite food, do *not* use a word such as "delicious": that is an emotional reaction and a judgment, not a perception.

Try to be as precise as you can. For example, "spicy" is a perception, but it is not as precise as "tastes of red chilis" or "the cayenne made my eyes water."

Try also to use as many senses as you can. Do *not* assume that one of your senses is not involved in the scene you are describing. For example, most sighted people tend to describe first in terms of what they see and often ignore other senses in describing. Further, because video conveys sight, sound, and movement, but no other sense perceptions, people used to watching video often ignore what they smell, taste, and feel on their skin when they describe. Try describing using *all* of your senses.

The table on the next page can help you complete the exercise. The goals of the exercise are to

- improve your sensory awareness,
- build your descriptive language, and
- help capture readers for your STEM writing by engaging them in what you describe.

We've filled in a scene name and a first line, "sight," of the table as a model, but you should choose your own scene or object and your own sense perceptions. You might also keep in mind that not everyone experiences senses in the same way or with the same intensity. As we think about ensuring reaching a wide audience, we want to keep that in mind.

Scene or object: **Rain begins to fall in the garden.**

Sense	Word	*More precise term*	*Phrase, sentence, or analogy*
Sight	Soil	wood chips, compost	The rain quickly turned the soil of wood chips and compost a deep brownish-black.
Smell			
Sound			
Motion			
Taste			
Touch/Texture			

Remember that this is just an exercise. It is not meant to be polished prose. Don't be afraid to experiment with language. If you are working with others on this exercise, see what each of you has come up with. Learn from each other.

Vivid Language vs. Impressive or Quantitative Language

Scientists tend to think of precision in quantitative terms. In Chapter One, pp. 27–29, we compared two articles on a single topic. One of these articles, from *Biological Conservation*, was numbers intensive, as one would expect from an IMRD journal, while the other, from the *New York Times*, listed a few statistics. Readers of IMRD journals expect lots of numerical data, especially in results, while readers of STEM journalism do not. In fact, too many numbers will cause most readers to lose interest quickly.

Nevertheless, **a few statistics can be a dramatic, vivid part of a well-written article.** In the Broad article on North Korean nuclear testing, the infographic on the explosive power of the four types of nuclear weapons uses four numbers in large print:

1 3 25 1,000

to show the great difference from one type to another. But these numbers are carefully chosen to create the greatest comparative impact—and therefore keep readers reading.

"Impressive" Language

Scientists, like professionals in other fields, are also trained to think that the best language is the most sophisticated and up to date in the research field. They learn this "insider" vocabulary from reading IMRD journals and from talking with peers and teachers, and they often feel stupid and unsophisticated if they don't know the most up-to-date jargon.

But while the "language of the discipline" impresses colleagues, it does two negative things to communication with non-specialists one is trying to reach:

- it intimidates, and
- it doesn't work toward reader understanding.

STEM writers need to give up both (1) heavy reliance on numbers and (2) "impressive" insider jargon if they want to reach non-specialist readers. The vivid language we've been describing in this section is intended not only to communicate vital messages to a broad readership, but also to show respect for the language with which your broad readership is likely to be more familiar.

Exercise 8G: Return to the article you chose for Exercises 8B, 8C, and 8D. Now, focus on the article's vocabulary. Especially look closely at words and phrases that are used to explain ideas or to describe objects or things that happen. If a goal of STEM journalism is to reach a broad audience with vivid language, has the writer succeeded?

Make a list of words, phrases, or sentences that you think are especially good at reaching this goal. Why do they succeed?

Also look to see how statistics are used. What purposes do they serve in the article?

On the other hand, do you think that the vocabulary could be improved? For example, are there terms that you think a non-specialist audience could *not* understand? Does any wording seem to be used to intimidate rather than inform? What wording might you substitute?

Vivid Writing through Analogy and Metaphor

Two other typical tools of STEM journalists are **analogy** and **metaphor**. Especially when trying to **explain processes** that are not perceptible without advanced instrumentation or that are theoretical, STEM writers rely on metaphor and other forms of comparison. Among the most famous of these is Einstein's example of the train moving at the speed of light (Einstein [1916] 2014), so that he might explain relativity. Another is the analogy of the power of nuclear reactions to pounds of TNT.

Among recent metaphors that have become popular to explain complex nano-processes is the use of the verb "cut" to explain the action of the enzyme CAS9 in the gene "editing" (another metaphor) process known as CRISPR (Ledford 2016). Like the examples above, the **purpose** of these metaphors or analogies is to **simplify** a complex process and to **relate what is strange and new** to something commonplace. Some metaphors, like "flow" for the movement of electrons in an electric "current," have been used for so long that we no longer think about what very different processes the terms refer to.

Visual Metaphors

In STEM journalism, multi-modal tools are especially useful in creating "visual metaphors" for readers' ease of understanding, as well as for highly vivid representations. In two of the articles offered as samples in this chapter, **drawings** have been used in this way:

- The Broad (2017) article on nuclear testing presents four small diagrams that colorfully and simply depict the proportions and arrays of "atomic fuel," "thermonuclear gas," and "solid thermonuclear fuel" in the four devices.
- The Bhattachargee (2024) article on stress shows colorfully (though not as simply) in its infographic a person in emotional pain to dramatize "The Chronic Stress Effect." The metaphorical image also draws attention to the lists of "Altered Systems," "Consequences," and "Stress Busters."

In both cases, the metaphor is not meant to be an accurate depiction, but serves to represent **comparative differences and similarities**. Likewise, charts and graphs work—if kept simple—as vivid visual representations of differences and similarities.

Exercise 8H: Use the two samples of STEM journalism that you used for Exercise 8E. This time, find examples of how the writer has used metaphor or analogy to represent vividly a nano-process, a theory (or hypothesis), or a process or phenomenon.

Find at least three metaphors or analogies in words in each article, and at least one visual metaphor or analogy in each.

How accurate do you find these worded and visual representations? Are they effective in making the writer's point? Try to imagine another metaphor or analogy that would work well in this article.

Ethical Challenges in Using Metaphors and Analogies

However, there are ethical challenges in using such metaphors. As we explored in Chapter Four, in simplifying there is always the **danger of over-simplifying**—thereby **misrepresenting** the difficulty of quality control in the scientific process and **exaggerating** the justifiable confidence in a new theory or method. There is also the potential danger that the metaphor works to mask concerns about the safety or unforeseen effects of the new processes.

STEM journalism, whether in magazines or newspapers, or in genres such as the research lab website, the professional organization newsletter, or the personal blog (see Chapter Nine), needs to balance the desire for the readers' ease of understanding with a concern for the accurate description of uncertainties and possible dangers. As stated in Chapter Four, there is often a fine line between **vivid writing** and **misleading advertising**—both STEM writers and STEM readers need to be aware of this need for balance.

That is why in this chapter we have coupled the advice for vivid writing with advice for creating an **evidence-based argument**. But we'd also suggest going beyond balancing. As you read STEM journalism and consider your own STEM articles for non-specialist audiences, look for explicit ways to counter a tone of enthusiasm for new knowledge, and for products based on it, with statements of caution. These, too, can be written with as much vividness and for as much ease of understanding as any other part of the STEM journalism article.

Exercise 8I: Again, use the two articles you used for Exercise 8H. Now look for ways in which the articles either

- include explicit statements that urge caution or provide contrasting viewpoints, *or*
- seem heavily weighted to a particular interpretation of evidence.

Do the metaphors and analogies used seem to you over-simplified and perhaps misrepresent a process or phenomenon, or is there ethical balance in the presentation?

Could you suggest improvements to each article in order to achieve balance without losing vividness?

CONCLUSION

We began this chapter by asking STEM students and professionals to consider becoming—to some extent—STEM journalists. For those who understand the burgeoning value of trying to reach readers beyond the lab, the IMRD journal, or the disciplinary conference, the chapter provides practical advice on many writing strategies: from taking on a different, more personal "voice" in writing, to using different forms of evidence, to organizing your writing in new ways, to experimenting with the goal of vividness in your prose. Lastly, the chapter asked you to think about and to enact the ethical implications of these shifts.

We have looked closely at several models of STEM journalist prose from large-circulation magazines and newspapers to see how their writers use these strategies and use varied multi-modal tools, including photographs and infographics. The many exercises of this chapter have asked you to apply the advice and the new techniques and, as with all the exercises in this book, to think of yourself as a designer and writer in the forms of writing you are studying.

The goal of the chapter was to help you gain the confidence to branch out in your writing, to think seriously of growing in your ability to communicate your research to new readers.

This chapter is actually only the first of several that will ask you to imagine yourself as a growing, improving STEM communicator, willing to experiment with new voices, styles, and language, and in new genres building huge followings in the age of the internet.

The next chapter, "Chapter Nine: Science Online—New Readers, New Voices, New Tools," will focus on increasingly common and popular genres in the digital space. As you will see, these tools offer writers an inexpensive platform to experiment in writing and to gradually build confidence in new forms of thinking about your research—as well as in reporting and reflecting on your research and its implications.

Exercise 8J: Return to the first exercise of this chapter, Exercise 8A. Now reread what you wrote there. How have you developed your thinking since that exercise?

As you contemplate writing your own STEM journalism, what tips from the chapter will you want to apply? From the examples of STEM journalism you've read and studied, what strategies do you want to adapt to your own prose?

CHAPTER NINE

SCIENCE ONLINE—NEW READERS, NEW VOICES, NEW TOOLS

Topics in this chapter:

I. Science Communication Online
II. A World of Blogs—Finding the Blog(s) for You
III. Studying the Major Types of Blogs
IV. Getting into Blogging for Yourself
V. Establishing Your Ethos
VI. Building Your Design

I. SCIENCE COMMUNICATION ONLINE

A lot of science communication happens in the lab, in the pages of peer reviewed journals, and nowadays online. Research articles, data repositories, etc., are all online and these are important for scientific research. We can also think of all the ways that scientists write online during their professional life, including emails, etc. What we mean by "science communication online" here includes more public activities, such as science blogs, microblogs, and videos.

Science communication online is not limited to what we can call internal and external genres, but offers ways to communicate science to a variety of audiences. When we speak of **internal genres**, we mean the kinds of **communication used during scientific research**, often by professional scientists. However, we should expand this definition to also include other people doing science, such as those involved with participatory research, citizen science, and so on. **External genres** mean those kinds of science **communication we commonly call popularizations**, where the outcomes of science are shared to inspire. Science communication online provides a space for many other genres, what are termed "parascientific genres."

Parascientific genres[1] are those genres that have some features of internal genres as well as external, but they are not strictly one or the other. For example, think about how the website Experiment.com structures its crowdfunding proposals. Proposals have some features that look a lot like what we might find in, for example, the abstract of a research article. Some parascientific genres are written by scientists, are deliberately written differently from peer-reviewed IMRD articles, and do not adhere to the same norms that the journals do. The style is more conversational, the vocabulary often less technically difficult, and the voice more personal, as we described in Chapter Eight.

The overall media architecture also shapes a genre. While the needs of the community are critical to shaping how a genre evolves to fit that community's needs, the tools we have to communicate are also important. With the rise of the internet, the ways to communicate through technologies changed. Those changes have antecedents, to be sure. Consider, for example, card index search systems at libraries and how those were precursors to technologies such as a web search. In studies of genre, scholars have traced how changes to technologies shape genres and genre systems, and these are important reminders as we face new technologies such as artificial intelligence (AI). If we know that there are some antecedents for a technology, we can take a longer view of how that technology might impact our communication practices. Let's consider a few key technology features that shape science communication online.

SEARCH TOOLS

You may have found one or more blogs or other types of websites that you turn to regularly. But if you have not gotten into blog reading and want to know where to begin, web searchers in almost any field have done some of the research for you. Just use your favorite search engine and type in, say, "chemistry blogs," or "engineering blogs," and you'll find sites that list—and often rank—a number of blogs in a specific area of interest.

---

1 Parascientific genres are theorized in Mehlenbacher and Miller (2017). You can also read about them in *Science Communication Online*, an open access book by Mehlenbacher (2019b): https://library.oapen.org/handle/20.500.12657/24833.

COMMENTS, PLEASE?

Another feature of blogs that sets most of them apart from either IMRD journals or most STEM journalism is their built-in interactivity. Most blogs include a comment feature, and the blog managers do want readers to comment. As you browse blogs, you'll see that many comment opportunities go unused—which usually indicates a lack of traffic on the blog. But there are also many blogs with many readers but few if any commenters. The great thing about "commentability" is that its presence tells readers that they are invited to add their ideas, but they don't have to.

PRESERVING THE SCIENTIFIC RECORD

Let's also consider how technology might challenge norms of scientific practice and its communication values. While the rapid speed of sharing is notable with online communication, a significant challenge has been preservation of materials. Such differences in communication technologies can fundamentally change the norms, values, and practices of communities. Canadian economistic and media scholar Harold Innis ([1950] 2022) wrote of such changes in his book *Empire and Communications*,[2] explaining how differences in communication technologies shape empires or civilizations. He looks across five millennia, but just in the ancient world we can see how communication technologies shaped those cultures, from the speed and portability of parchment in the Roman Empire, which reached across a vast area of land, to the enduring technologies of stone inscription of ancient Egypt. For science, we can trace a long history—don't worry, not in this chapter—of how technologies have shaped its practice. Online technologies are changing key norms around preservation. Even in revising this book, we found that some of the links to social media videos or websites were already (in a short five or so years) gone, and we could not find new links to those materials. For the scientific record, this disappearance is highly concerning, and many people, including scientists, are working to develop new ways to ensure that we archive science communications online.

If, for example, some federal government databases on which the scientific community and even the larger public have come to rely are taken offline, groups of scientists and scientific organizations would need to archive documents to keep them available. So as you write into online venues, consider both their staying power as well as the challenges to preservation.

---

2 Notably, Innis's daughter, Dr. Anne Innis Dagg, was a leading zoologist studying giraffes—sometimes called the Jane Goodall of giraffes—and made important contributions in communicating this science to researchers, broad publics, and children: https://anneinnisdaggfoundation.org/.

II. A WORLD OF BLOGS—FINDING THE BLOG(S) FOR YOU

Let's consider one of the original and important genres of online science communication: the blog. Blogs are a relatively long form of science communication online, when compared to microblogging or videos, for instance. Blogs also offer an excellent example of parascientific genres. Some blog forms tend to be oriented toward not only a broad public audience, but at other experts. Science blogs have become prolific in all fields and have become a common means by which new ideas—and new voices—in science have been recognized. These blogs are both **stand-alone sites** by individuals and research teams, or **interactive adjunct sites** to established companies, journals, and agencies. New styles, including freer use of first person and multimodal capabilities, have made blogs experimental sites for science communication.

Frequently, the blogger writes in the first person, and the purpose of the blog entry is explicitly to express a point of view about an issue or phenomenon. Often the tone is humorous or even angry, maybe sarcastic, certainly opinionated. There is nothing in most blog entries that emulates the IMRD format, especially not the deeply detailed descriptions of methods and results of the IMRD form. However, some blogs do venture beyond what we would think of as popularization, including discussion of methods or problems with methods, for example.

Yet, at the same time, our students know that blogs are not like most science journalism (see Chapter Eight). First, the journalist is usually not a professional scientist, but a professional writer who is primarily making science compelling and exciting to wide audiences. The journalist summarizes the peer-reviewed research and often interviews researchers or other experts.

In contrast, the science blogger usually proclaims his or her professional ethos in a biographical statement that highlights their credentials. Whereas science journalists usually present themselves to their readers as interested outsiders who rely on scientific experts to verify what their articles describe, science bloggers feel empowered to speak for themselves—even though they also often quote other researchers as further evidence for their points of view.

Therefore, the science blog holds a kind of middle space between the credibility of the IMRD genre and the broad readability of STEM journalism:

- The blog as a genre offers writers the freedom to experiment with **voice and tone**, length and stylistic features, and it doesn't require writers to go through the strict procedures of peer review. That freedom, though, means that readers often don't know what to expect, and that leads to reader uncertainty about a blog's value, even while the idea of freedom is attractive to potential writers.

- Meanwhile, the professional ethos of the writer can give the blog a scientific prestige that most science journalism does not have, and that can make the blog credible as science.

Blogs can be a useful avenue for would-be writers to practice writing in public. But this interactivity also signals the capability of the internet to encourage the building of communities of interest. If you already like certain blogs, and have entertained the idea of perhaps starting your own blog, a good way to get your feet wet in blogging is to write comments to pieces in your favorite blogs. (See Section IV—"Getting into Blogging for Yourself"—for more on getting started.)

Finding such popular sites will give you a good idea of writing styles, content, media, and layout that have found large audiences. We're not implying that your initial goal in blogging should be to join these ranks of the most read, but it won't hurt you to see the factors that attract large readerships. Popularity rankings and searches through Facebook, Google, or Yahoo are not the only ways to find blogs that may suit your interests. The low-cost web platforms that allow you to create blogs fairly easily (described in Section IV) enable keyword searches that can take you to other blogs that they sponsor and that are in your interest area.

For example, if you begin to design a website or blog on wordpress.com, you can search their entire database of billions of blog posts to find

- terms from your field, and
- other bloggers in your area of interest.

Finding these fellow chemists, engineers, botanists, neuroscientists, geologists, etc. may be the best way to get stylistic ideas for your own blogging—in addition to giving you access to great information. You can let some of these sites inspire you, but don't let them intimidate you. There's room in the blogosphere for first-time bloggers like you, and time for you to build up to the expertise of the veterans.

You'll also find through keyword searches many uses of your keywords in blogs that have little or nothing to do with your interest area. However, these "discoveries" can be illuminating as well, as you'll see how writers with very different interests and styles use what you may have thought of as "your" language.

III. STUDYING THE MAJOR TYPES OF BLOGS

The two major types of blogs are

A. stand-alone sites, and
B. sites that are part of organizations and publications.

Both types can be really instructive to read for fun *and* to study in terms of their **rhetorical features**, including

- their purposes,
- their intended readers,
- their genres,
- their ethos/credibility,
- their voices,
- their tones,
- their argumentation, including evidence,
- their organization and navigation, and
- their uses of visual and aural media.

But if you are planning to start your own blog, you'll want to pay particular attention to the stand-alone blogs. Why? Because the stand-alones have to create their own ethos. They can't rely on the larger "parent" organization or publication to give them instant credibility. Nevertheless, all blogs can be sources of techniques that you may want to remember and try out.

A. ANALYZING A SAMPLE STAND-ALONE BLOG: "LIFE IS SHORT, BUT SNAKES ARE LONG"

Managed and written by then-PhD student Andrew Durso, "Life Is Short, but Snakes Are Long" (https://snakesarelong.blogspot.com/) was created on Blogspot in 2012, and announces just below the title that the site presents "Snake biology for everyone." This ambitious **purpose and sense of audience** is reinforced by a varied **style** that is sometimes highly technical, yet is mostly accessible to a very broad audience that knows little about snakes and their behavior.

The August 31, 2017, post—Durso's 96th—is a case in point. Titled "How Many Snakes Are Venomous and How Many Are Constrictors?" (Durso 2017), it begins with clear, non-technical language intended for a vast readership:

> Many people are aware that some snakes constrict their prey, and others use venom to kill their prey. Recently, somebody asked me what the breakdown was, and I had to admit that I didn't know exactly. (Durso 2017)

Durso proceeds to give his "initial estimate," but then relates his own search for the best answer. In so doing, he shares citations from a broad range of **credible sources**, but always in a style that is aware of his non-specialist readers:

> I found that the answer to this question is not as simple as it may seem. Many snakes unambiguously use venom or constriction, but many use neither, and some use both! Of course the data are not as detailed or abundant as we would like. What follows is a breakdown of the categories I used, and some interesting exceptions that I uncovered. (Durso 2017)

Durso then organizes his information so that any reader can easily follow. Though there are some longish paragraphs, he uses many bullet points to allow lists of characteristics to stand out, and he uses the bold subheads

- "Constrictors,"
- "Venom,"
- "Neither,"
- "Both," and
- "Unknown" (Durso 2017)

to allow the reader to easily navigate the whole piece.

To build his ethos in the blog, Durso uses a variety of methods and tools:

- His bio is located on the right side of the screen for each of his posts.
- The bio is followed by links to his Twitter (now X) feeds, his Google Scholar profile, his ResearchGate profile, his Academia.edu profile, and his complete CV. Interested readers can see his peer-reviewed publications, grants, and other evidence of his credibility as a scientist.
- Each entry in the blog is followed, in IMRD article fashion, by a bibliography of sources cited—plus a link to the list of all sources read in preparation for the entry.
- The vocabulary is varied so that more technical terms and Latin names are also included throughout the post—but never in a way that makes

the explanations unclear to a non-specialist reader—just as we might expect in a parascientific genre example that is trying to reach multiple audiences. By splitting his audience in this way, Durso is able to satisfy the desire of his specialist readers for a more technical explanation and the desire of his broader audience ("everyone") for a less-technical explanation.

This last point about style is worth elaborating. Here's an example of how his **audience splitting** works.

In the "Both" section of his entry, in which he describes snakes that use both venom and constriction, he writes the following:

> A review by Rick Shine & Terry Schwaner brought together data on numerous Australian elapids that, although they clearly have and use venom, also use their coils to subdue and hold prey while envenoming it. In many of these species, including tiger snakes (*Notechis*), brown snakes (*Pseudonaja*), curl/myall snakes (*Suta*), whip snakes (*Demansia*), Australian coral snakes (*Simoselaps*), crowned snakes (*Cacophis*), and olive seasnakes (*Aipysurus laevis*), the coils are not used alone as the primary method of prey subjugation, and one recent paper suggested that we think of them as "*part of a 'combined arsenal' of prey subjugation strategies*." (Durso 2017)

Notice that for his fellow researchers, Durso includes (1) **links** to two peer-reviewed papers and (2) the **scientific names** of several snake species. However, the text remains accessible for non-specialists, because he paraphrases the main point of each article in nontechnical terms. In effect, within this one paragraph, the post addresses the needs of very different readerships—through a quite simple strategy.

Tone and Voice. Notice also how the blogger varies his tone and voice in order to engage his diverse readers and keep them reading. The title of the blog, "Life Is Short, but Snakes Are Long," puns on the two meanings of the word "long" (time and size) and so creates a humorous tone that will engage a broad readership. This non-serious tone continues in the first paragraph of the post, when he recalls—in a self-effacing way—his not being able to answer the question that someone put to him. So, he creates a mood of curiosity that engages the reader in his quest for an answer. We are hooked—and can't wait to hear more from this curious voice.

Then, as the post goes on, Durso's voice becomes more scholarly, as he cites sources and describes the habits of different snakes. As we read on, we have no doubt that he knows what he's talking about. But his voice never loses that curiosity. And, because his topic is how snakes kill their prey—he always keeps his readers on the edge of their seats with a bit of fear. In the very last line of the post, he makes us wonder how we are like the snake and its prey, as he asks us to think about "the benefits of that defining snake trait: being able to consume prey almost as large, and sometimes much larger, than yourself!" (Durso 2017).

So, by varying his tone between the humorous, the curious, and the serious, and by varying his voice between the personal and the scholarly, he reaches his diverse readers, builds his ethos, and keeps the reader engaged. Notice how similar these uses of tone and voice are to the techniques of proficient STEM journalists, as we saw in Chapter Eight.

Use of Graphics. In the sample we've been analyzing, **graphic presentation** is used in several ways that reinforce Durso's organization of the post and his ability to engage his readers:

- By using **short paragraphs** and **bullet points**, the blogger helps the reader focus attention on each important point. Ideas don't get lost in the middle of dense paragraphs.
- Every post is **formatted similarly**, so that readers can be comfortable with the layout.
- Each post is surrounded by the same information in the same locations on the page. For example, Durso's bio and links to his profiles are always on the right side of the page. This uniformity promotes **ease of navigation** of the entire site.
- Similarly, use of **bolded subheads** keeps the reader focused on the progression of the article from section to section.
- Many **links**, in orange, are embedded in paragraphs. These attract the reader's attention. Besides, by using links to other sources, the blogger doesn't clog his own site with lengthy paraphrases or quotations that can slow down the reader. The links allow readers to choose if they want further information.
- The writer is sparing in the use of graphs, charts, and photographs, but **chooses carefully so that the graphic devices achieve major impact**.

In this sample post on how snakes kill their prey, there are only seven illustrations, each reinforcing a main idea in the post:

- a pie chart in the opening paragraph showing the proportions of each types of killing practice (constriction, venom, etc.)

- a drawing of how one snake (*Anilius scytale*) constricts its prey
- a photo of a venomous snake, a black mamba, "eating a bird"
- a photo of *Dipsas indica* "coiling around a snail"
- a photo of *Pseudonaja textilis* "constricting a mouse"
- a photo of *Elaphe quadrivirgata* "constricting a frog"
- a diagram showing the evolution of the different forms of snakes killing prey

Clicking on each graphic enlarges it, but, unenlarged the graphics don't overwhelm the text, while still being clearly visible. As with the use of links, the use of graphics allows readers to choose when and where they want more detail or information. In this way, the blog is tailored to the needs of different types of readers.

Exercise 9A: Find a stand-alone STEM blog of interest to you. Reflect on how you found the blog. Did you use a general web search tool, did you know of one already, did you search a blog website, or maybe you used AI to create a list of blogs for you (bonus reflection question: were all the blogs the AI output listed real?).

Once you've identified your blog and read through it a bit, use the list of rhetorical features on p. 230 to analyze how the blogger has constructed a blog post to reach readers. Be sure to take notes on all eight features. The analysis just above of "Life Is Short, but Snakes Are Long" (Durso 2017) can be a model for how you perform your analysis.

What aspects of this blog are effective in achieving reader interest? How has the blogger created an ethos of credibility?

What suggestions might you have to redesign the blog to enhance effectiveness? If you are doing this exercise with others also looking at this blog, compare your observations and suggestions.

B. ANALYZING BLOG SITES THAT ARE PART OF ORGANIZATIONS OR PUBLICATIONS

Though the great majority of blogs are written and self-sponsored by individuals, there are also blogs sponsored by publications and organizations. Sometimes, an organization may sponsor several blogs, as in the case of PLOS (Public Library of Science), which sponsors blogs written by PLOS staff in different research fields and independent blogs hosted by PLOS that are written by scientists and journalists independent of the PLOS staff. As explained on the PLOS main website:

> The shared mission of these contributors is to promote greater understanding of breakthrough science for a variety of reader types, including policy makers, the academic science community, researchers, medical and mental health practitioners, journalists and the general public. (PLOS n.d.)

Organization-sponsored blogs may be managed by publications, such as *PLOS One* and *National Geographic*, by universities or university departments, by professional societies, or by commercial sites (e.g., Engineering.com). How these blogs serve the desires of the organization is represented by this statement on the PLOS site:

> Beginning with the launch of the organization's main blog, plos.org, now known as The Official PLOS Blog, back in 2006, PLOS quickly realized how informal communication can catch PLOS authors and readers' attention and help share and explain important scientific ideas. (PLOS n.d.)

As well as being "informal communication" that "can catch PLOS authors and readers' attention," the organizational blog encourages reader interaction through comments (PLOS n.d.). This kind of give-and-take is meant to build a sense of community that is difficult to achieve in the serious formal atmosphere of the peer-reviewed journal.

But the give-and-take on PLOS blogs is held to strict "community guidelines," 12 in total, including "PLOS BLOGS reserves the right to reject, at our discretion, any comment that is insufficiently supported by scientific evidence, is not constructive, or is not relevant to the original blog post" (PLOS n.d.).

This policy raises questions about potential commenters who worry that the journal will feel that their comments are "insufficiently supported by scientific evidence." Indeed, the guideline shows that the intended readership of the PLOS BLOGS Network is probably narrower than that we saw in the sample stand-alone "snakes" blog. Consider which groups of readers would actually feel qualified to "sufficiently support" their comments "with scientific evidence." It may be that people with important lived experience but without scientific evidence for their claims, for instance, would feel excluded. This effect on potential commenters may be compounded by the penultimate guideline, which states, "PLOS BLOGS reserve the right to remove any content that violates any of these guidelines, to block repeat and/or egregious violators from posting, and to suspend accounts as we deem necessary" (PLOS n.d.).

While it is important to foster a culture where many voices are included, this last item gestures toward the other policies that include not defaming or calling someone names or otherwise attacking, etc. Science has long been an exemplar for argument and debate, and that's not what is being discouraged here, but rather the **vituperative nature of online discourse**. Such guidelines are a reminder of how the intersection of professional scientific norms and the norms of online discourse are merged in parascientific genres such as blog posts.

As we examine a sample blog entry in PLOS, we can look for other indications of the breadth or narrowness of the intended readership. As you consider the breadth of readership that you would like for your own blog, the types of policies and guidelines you announce will be part of what readers see that will determine if they feel a part of your blog's community.

As science is also an international enterprise, it is also important to think about the norms, values, and, indeed, sometimes laws of nations. Not every country, for example, has the same understanding of free expression, and prohibitions of certain forms of hate speech, for instance, may govern online communications.

Looking Rhetorically at a Sample Entry from the PLOS-Sponsored Blog "DNA Science"

If you browse the many blogs in the PLOS BLOGS Network, you'll find a wide array of styles and voices, and of course topics. We recommend that if you spend some time doing such "rhetorical browsing," you can determine what you'd like to achieve in your blog and pick up some provocative ideas for blog design. Remember that all PLOS entries and blogs have been judged acceptable within the PLOS "community guidelines" cited above.

The blog "DNA Science" (https://dnascience.plos.org/), managed and written by geneticist-journalist Ricki Lewis, is among the PLOS-sponsored blogs not written by PLOS staff. We chose it for this brief analysis because from entry to entry it displays Lewis's voice (just as the "Snakes" blog highlights Andrew Durso's), so from entry to entry we get a good sense of how the blogger tackles different topics and manages the rhetorical features.

The post from October 5, 2017, "Brain Cancer in Kids: Tailoring Treatment Based on Mutations" (Lewis 2017), shows Lewis delving into a topic of great interest to a broad array of audiences. Let's do a brief analysis via the rhetorical categories to see how she handles the topic.

Purposes. The second part of the title, "Tailoring Treatment Based on Mutations," may not mean much to non-specialists, but it becomes clearer as the essay goes on. Lewis's narrative begins by comparing TV ads for two cancer drugs, Keytruda and Opdiva, with focus on how each manufacturer chose subjects for the clinical trials. Gradually the point of the comparison emerges, as she points out that one of the companies chose subjects based on the presence of a particular protein, PD-L1, in subjects' cancer cells. Knowing the proteins, she states, leads to better predictions and more accurate treatment strategies: "The protein topography of cancer cells, determined by their genes, is so important in selecting treatment that it's now considered along with or even before body part" (Lewis 2017).

This main point doesn't come until a third of the way into the 1,200-word essay, and even before she actually addresses the topic of "brain cancer in kids." But once she does move to this topic, Lewis uses her description of protein detection in genes—the first part of the essay—to make a firm statement about childhood cancers: "Brain tumor cells from an adult and a child that appear the same histologically often in fact harbor different mutations" (Lewis 2017).

Showing how this is true, and what it means for diagnosis and treatment, is the most important **purpose** of the essay.

Intended Readers. Lewis captures a **broad readership** with her first sentence: "I'll admit it. I was sucked in" (Lewis 2017). She goes on to describe the scenarios and main characters—"Sharon," "Donna," and "an older gent"—of two competing TV ads for popular cancer drugs. She relies on reader knowledge of the ads and products to gain readers' attention. Her intended audiences clearly include both non-specialists and medical professionals.

As the essay becomes more technical, however, it becomes **less directed at non-specialists**, such as parents of children, and more readable by clinicians, researchers, and those familiar with pharmaceutical decision making. It doesn't really return to the language of the first third of the essay, except in a couple of clauses mentioned near the end of the essay.

Ethos/Credibility. Lewis's brief bio (plus photo), which appears in small font at the end of the post, describes her credentials as both geneticist and STEM journalist. If you click on the heading "About This Blog," featured on all posts, you also get this information. In the essay itself, she refers, with a **link**, to one of her own publications in the **scientific literature**.

Moreover, the entire blog has credibility as part of the PLOS BLOGS Network. Note, though, that, as in most STEM journalism in newspapers, there is no end list of sources, and only a few links in the essay to peer-reviewed sources. A large portion of the ethos of this essay comes from the association with the PLOS Network.

Voice(s). The personal, vulnerable, self-effacing voice in the first sentence attracts the broad readership, but that voice disappears after that point, except for a one-clause reference to her family struggles with cancer: "as my mother and father-in-law did" (Lewis 2017). The voice that takes over for the rest of the essay is the self-assured voice of the genetics lecturer and experienced journalist who can delve into the diagnostic issues, paraphrase key pieces of current research, and confidently reaffirm her main point. She does bring back, at the very end of the essay, the TV characters she introduced at the start, but only in a final clause, "so that future cancer patients can see many more tomorrows—like Donna and Sharon" (Lewis 2017).

Tone. As with voice, the tone at the outset is distinct from the tone thereafter. The beginning tone seems forthright and humble, willing to admit doubt: "I'll admit it. I was sucked in." But the tone after that point is very confident, as the blogger systematically moves through the stages of the argument about "protein topography" and its benefits for diagnosis and treatment (Lewis 2017).

Argumentation, Including Evidence. The argument of the essay is to convince readers of the two main points described above under "Purposes." The evidence to support the argument comes from a number of sources:

- Lewis's close reading of the two TV ads
- Lewis's own IMRD article (linked) from *Lancet Oncology*
- an IMRD article (linked) from the *Journal of the Advanced Practitioner of Oncology*
- an IMRD article (linked) from *Cancer Cell*
- a quotation from her interview with the principal investigator of "the glioma team at The Institute of Cancer Research in the UK"
- a quoted statement (no source given) from the World Health Organization

In her descriptions of research, it is not easy to tell which sources she is citing at any given point, because few quotations and no other precise citations are included in the essay. In order to convince her readership, the author appears to be relying substantially on the **citations** she does have, as well as on her confident **tone** and her **ethos** as geneticist and journalist. It should be noted that she does not refer to her points of view as controversial; nor does she cite or paraphrase studies that might run counter to her direct line of reasoning.

Organization and Navigation. The title sets the focus of the essay on "treatment" for "kids with cancer." Readers will be looking for details about this focus, even though the first part of the essay deals with a different topic. Another organizing feature is the **subheads**, which are in all-capital letters but not bolded, and which are set off from the rest of the text. **Bullet points** are used twice in the essay: to list ways that normal cells "go haywire" in cancer, and to compare three groups of mutations in childhood gliomas, according to differences in "median survival rate." In both cases, the language of these paragraphs seems intended for professionals, not for a broader readership.

Uses of Visual Media. There are five visual images in the essay, which add color to the blog but are not obviously pertinent to the argument:

- a large color drawing of a silhouette of a person, with the words "Melanoma" and "Breast" in the drawing
- a color photo of a section of brain tissue, with the caption "Glioblastoma"
- a color photo of President Obama "greet[ing] glioma patient Jack Hoffman" (Lewis 2017)
- a color drawing of what might be meant to represent a DNA string, but which is not captioned
- a color drawing of a medial cross-section of a child's brain, with parts of the brain marked

Three of the images do relate to the general topic of childhood brain cancer, and may be meant by the blogger to attract readers interested in the topic. But the visuals are not meant to clarify or illustrate her argument.

Exercise 9B: Find a blog of interest to you that is sponsored by a publication or an organization. Use the list of rhetorical features on p. 230 to analyze how the blogger has constructed a blog post to reach readers. Be sure to take notes on all eight features. Our analysis just above, of the entry "Brain Cancer in Kids: Tailoring Treatment Based on Mutations" from the blog "DNA Science" (Lewis 2017), can be a model for how to do your analysis.

- What aspects of the blog you've chosen are effective in achieving reader interest?
- How has the blogger created an ethos of credibility?
- How much does the blogger depend on the ethos of the publication or organization for the effectiveness of the entry?

What suggestions might you have for how the blog could be redesigned to enhance effectiveness? If you are doing this exercise with others also looking at this blog, compare your observations and suggestions.

IV. GETTING INTO BLOGGING FOR YOURSELF

Finding a focus for your blog. Bloggers create their writings for all sorts of purposes and in different ways. Of course, you might intend your blog to become the definitive source, the go-to place, for your research area. That's a great ambition, and you'll find through the search techniques noted above (pp. 228–29) some outstanding examples that you might wish to use as models.

No pressure to be great. The wonderful thing about the blog process is that there's no pressure on you to be great right away—or at all—and you can focus your new STEM blog as you'd like to. In the days before the internet, many of the people who now blog kept paper journals, diaries, or scrapbooks/albums. You can keep the blog as a place for daily or weekly jottings, notes, or thoughts intended for yourself only. Remember that you don't need to publish your blog until you are ready, and you can always keep parts of it private while publishing other parts. You can also take down pieces you've published or edit them as your ideas change. You may wish, however, to note those changes in order to preserve the scientific record online.

Frequency? Once you do find a focus that you want to continue, you can post entries as often as you wish. While some bloggers post every week or month, or more often, many post regularly for a time, then are silent for long periods. Some readers find it annoying when a blog they really like is not followed up. When thinking about post frequency for blogs and other online platforms, it is important to understand audience expectations in order to ensure consistency and grow your audience.

How about comments? Blogging platforms are usually designed to give you the option of receiving comments. They also allow you to receive and judge the comments before you allow them to be published as part of the blog. When you set up your blog, you can include guidelines for commenters—such as the "community guidelines" we quoted from the PLOS BLOGS Network. The guidelines you post will help determine the nature of the "community" that your blog creates, or if you really want a contributing community.

Patience. Because the number of bloggers is in the millions, and even the number within single STEM disciplines can be in the hundreds or thousands, don't expect your stand-alone blog to become an overnight sensation—if that is your goal. When you plan to start your blog, do some thinking and writing about your ambitions for the blog, much as you would think about your ambitions for your research or for your career as a whole. Not all bloggers want to be known for their blogs. Many really do treat the blog as an archive of experiences or a seedbed for ideas they will develop in other ways—in their IMRD research, for example. Others are happy with a small readership of folks with similar interests.

For the ambitious. If you do plan for your STEM blog to help make you famous or at least to advance your career, it's helpful not to think of the blog as an entity unto itself, but rather as one part of your overall résumé. Both bloggers we studied in the previous section, Ricki Lewis (2017) and Andrew Durso (2017), built their ethos with IMRD publications. Lewis is also a prolific STEM journalist. Some bloggers also contribute to their accounts on Instagram or other **social media platforms**, and advertise their blogs in those forums.

So, consider how the blog you're planning can add a further dimension to your scholarship, your writing, or your life:

- Do you see it enabling you to reach a wider audience for your research, as you shift voice and tone—and perhaps use some of the features of STEM journalism that we studied in Chapter Eight?
- Could it allow you to give breadth to your research interests, if it focuses on a somewhat different aspect of your research discipline than what you concentrate on in your IMRD work?
- Do you see it helping to focus your community involvement, if, say, it provides a forum for you and others to engage in community projects, whether they relate to your research area or not?

Whatever role you imagine your blog playing, thinking carefully and writing about your ambitions for the blog will help you in the next two important aspects of getting into blogging:

- building your ethos, and
- designing your blog.

We'll explore these elements in the next two sections of this chapter.

V. ESTABLISHING YOUR ETHOS

The sample analyses, A and B, in Section III above, show how two prolific STEM bloggers both build the ethos that they bring to their blogs and cultivate that ethos through their blog sites and posts. Thus, we can say that their blogs both depend on the ethos that they bring and then add to their ethos through the quality of their blog posts.

As you think about the blog that you will create and maintain, keep in mind these two ways of establishing ethos:

- your ethos based on your prior credentials and achievements, and
- how the blog can enhance your ethos through the quality of what you write and how you design the blog.

Let's consider each of these types.

PRIOR CREDENTIALS AND ACHIEVEMENTS

In both analyses of sample blogs, we spent time looking at the design elements that each blogger uses to announce prior credentials and achievements. Study these again in Section III. Note that both bloggers keep their brief **biographies** easily viewable by their readers, either beside or at the bottom of each post. Durso (2017) augments his easily seen bio with links to his CV and other academic profiles. Both bloggers also refer to their IMRD publications in their entries. In short, both make a clear effort to keep who they are and what they have accomplished before the reader. Consider how important it will be to you and to the credibility of your blog to keep credentials and achievements easily seen by your intended readers. How you answer this important question will depend on your purposes for the blog and the relationship you wish to forge with your intended readers.

For example, if you want your blog to enhance your research career by bringing the results of your research to a wider audience than that reached by your IMRD publications, then you'll want to keep your professional credentials and IMRD record easily available to your blog readership. This is not bragging, as long as you stick to the facts that will help your new readers respect the background that you bring to the blog.

Don't forget that STEM blogs are ambiguous in terms of their respectability as science, so anything you can do to assure your readers of your credibility is vital to the success of the blog.

On the other hand, if your purpose is modest—say, to archive your experiences and observations in the topic or activity that your blog addresses—and if you want to reach only a small readership of friends, relatives, colleagues, and internet visitors, then it may be unnecessary, or even counterproductive, to parade your credentials. As always in rhetoric, keep thinking about your purposes and audiences, and don't be afraid to adjust what you say and show about yourself because these two elements might change.

ENHANCING YOUR ETHOS THROUGH THE QUALITY OF YOUR WRITING AND DESIGN

No matter how impressive your credentials and your prior achievements might be, the quality of the writing, content, and design of your blog will be essential to its ethos. Readers will come back looking for more posts, and are likely to sign up as "followers" only if what you write about and how you present it convinces them that reading your blog is informative, thought-provoking, and entertaining.

Especially if your purpose in blogging is to enhance your reputation and build your career, you want to be sure that the blog will be a positive force and will not, instead, drag down the ethos that you have so carefully built. We don't say this to alarm you, but like anything else that you do to develop your scholarly or research career, publishing your blog and being ambitious about getting attention for it can work for you or against you, depending on the quality of your work.

In the next section, the last one of the chapter, we'll look once more at the rhetorical features—and how you can use them to create that positive force for your ethos.

Getting Feedback on Your Blog

Seek competent, truthful feedback on your blog from readers you trust. This is not the same as allowing readers to make publishable comments on the blog, though that can be part of the feedback process. Such comments are likely to be about the value or accuracy of information in specific posts, which can of course be helpful. But it's hard to know how much to trust those comments. The feedback we mean here is the systematic asking of views from your trusted reader or readers in terms of your design, your voice, your topics—or any other feature of your blog about which you want competent, thoughtful responses, and advice.

As with the feedback processes described in earlier chapters, create a list of questions you have about the specific features and choices you've made. For example, you can use the eight rhetorical features from p. 230 to craft questions for your trusted reader:

- From your reading of my blog so far, what do you think my main purposes are for the blog? Any suggestions?
- What kinds of readers do you think would be interested in my posts? Is there any other type of reader you think I should be addressing?
- How would you describe the tone of my posts? Do you see me shifting tone? Any suggestions?
- And so forth.

Ask your trusted reader or readers to give you some time to talk thoughtfully with you about the blog—and assure them that you value their honesty and their insight. More than likely, they will ask you about your goals and hopes for the blog. Be as honest with them as you are hoping they will be with you.

VI. BUILDING YOUR DESIGN

Exercise 9C: Begin your design process at this point by writing for yourself about how you'd like your blog to represent you and present your topic. If you're not planning a blog, you might think about how to do this for a platform you use, such as Instagram or TikTok. As you've gone through this chapter, you'll have become more adept at not only reading blogs and at thinking about how you'd like your own blog to look, but also at thinking about how to use online tools to effectively communicate science.

As you've studied the sample blogs and other blogs in the exercises, you'll have picked up ideas for the design of online science communication, as well as seen design choices that you want to be sure to avoid!

So now write about which ideas you think work and which don't. You may be writing about a totally different platform where the norms of blogs wouldn't work. Why? Explain how you'd design your science blog, Instagram account, or other platform to communicate about the science you love. This pre-writing will help you make choices among the many options you'll have when thinking about communicating science online.

Depending on which tool you decide to use, you need to make some design decisions. Blogs require a lot of decisions, whereas social media platforms might be more restricted in terms of design. In either case, you should think about design elements, including those that perhaps aren't already built into the tool you're using. For example, if you're using a microblogging tool or video site, can you include video or image captions to ensure your content is accessible?

Let's consider blogs again to think through some of the decisions (but these could be adapted to the platform you wish to use).

HOSTING YOUR BLOG

Use your favorite search engine to find blogging platforms, of which there are many. These vary in price (some advertise as being free) and that price is often itemized according to the services they can provide. If you are fortunate enough to belong to an organization that can provide its own hosting services for members, you may be able to get free web hosting and the ability to use the organization's internet domain name. Be sure to find out about any restrictions in terms of design options or content that may be part of using your organization's tools and services.

Design Options On Hosting Sites

Once you start browsing hosting platforms, you'll likely find several options for the basic layout of your site, and it's a good idea to try out several, even all, of these to see how they might work in terms of how you want your site to look and "feel." How a site feels to a reader will include the following:

- the ease of navigation (links within the site and to other sources)
- the ease of reading the text (size and clarity of the font)
- the ease and pleasure of scanning a screen to find what you want
- the ease of understanding graphic elements and finding pleasure in the colors, images, and sounds that these graphic elements can provide
- the compliance with accessible design principles

Most first-time bloggers will choose a relatively simple design that will be easy for them to work with. Then, once they have identified how the choice falls short of their ambitions, they go back to re-design.

SOME SPECIFIC DESIGN CHOICES

For the new blogger, here are some features you should be sure to attend to as you design the look and feel of your site.

Navigation. How can readers move from place to place in the blog? Make sure your links work and keep working (this means regular checking of links as you maintain the site). What search function is on your site? How easy is it for readers to use?

Organization. How easily can new readers understand the purposes of the blog and the kinds of information they are likely to find there? (Most blogs have an "About" link that will give readers this vital information.) In each entry, how clearly does the writer introduce the purposes and content of the entry? Do titles help readers know what's coming? Are there helpful subheads as the entry proceeds?

Balance of text, images, and links. Web design theory keeps changing as tools evolve, and every blogger will see the idea of balance somewhat differently, depending on the blog's purposes and intended readers. If you really want your readers to pay attention to what is in your blog entry, don't so clutter the page with links to other sites that readers will navigate far away from your blog. Likewise, if your blog is not primarily a photo site, keep images small and not so numerous that they overwhelm the written text (but as images probably do increase engagement, choosing them carefully is important). As you browse sites, look for how the blogger achieves (or fails to achieve) balance so that emphasis stays where it belongs.

"Readability" of your writing. This refers partly to organization and to the reader's ability and ease in following the flow of ideas in your entries (see "Organization," above). But it also refers to the appropriateness of the wording you use in your entries. In Chapter Eight, we looked at tools and techniques that STEM journalists use to engage non-specialist readers (p. 201). Review these if you need a refresher.

The main point here is that bloggers, especially those—including STEM specialists—who spend most of their time communicating with a special-interest group of other researchers, can forget they are trying to reach non-specialist readers. It's amazing how easily we can get so in love with our technical train of thought that we forget to keep our passengers aboard as we roll along.

Paragraphing and white space. Related to both navigation and readability, be aware of how bloggers need to limit the length of paragraphs and use white space to focus reader attention. Sure, some bloggers will want readers who love to wade into dense forests (swamps?) of words; some readers don't care if they get lost there. But not most readers.

A simple technique to try is this: as you review your draft of each entry before you publish it, examine the appearance of the text on the screen. Every time you see a long paragraph (more than a couple of inches on the screen), go into the thicket and find at least one place to break the paragraph. You'll find that this is easier than you might think. Most readers, including you, will appreciate the sharpening of focus that comes from having a manageable chunk of text to digest before moving on to the next one.

Algorithms and Tags. But wait, you might say: not all "readers" are human. This is true, and how we think of the readers part of search engine optimization is another important factor. Optimizing for algorithmic readers is going to vary from the time we write these words and across platforms. It is an important consideration in getting a reader/viewership and part of how you will think about the content you develop on your blog or even in the video content you're creating and sharing on platforms. One way that some blog platforms encourage their users is by **enabling tags** that will alert algorithms to send traffic to that blog.

CONCLUSION

As online forms of science communication continue to evolve alongside the needs of scientists and other audiences, as well as along with new technological innovations, researchers will need to continue to keep apprised of these new forms. In this chapter, two major types of STEM blogging provided an example of how you can communicate science online. Stand-alone sites and those affiliated with organizations or journal publications provided a basis for analysis. These analyses laid the foundation for advice to new STEM bloggers in getting started; but as we noted, you can adapt many of these strategies to other online forms.

It remains to be seen how AI will impact science communication online, although elsewhere we have discussed challenges AI poses to science communication. It is likely that in the coming years a proliferation of forms will include aspects of AI, as well as increasing use of video forms, and this space too will have to contend with ethical issues we have previously discussed, including the challenge of misinformation but also questions of data privacy and security.

From a communication perspective, STEM professionals might observe how these bloggers and their followers establish ethos. They might observe how these writers/designers use language in ways that other members of these communities not only understand but that makes them feel welcome to participate. From these respectful observations—and perhaps participation in some of these communities—STEM professionals can increase their own ability to write effectively for a wide range of readers.

Chapter Ten moves from the blog to a different popular form of multimedia STEM communication: the poster and its very recent digital/online spinoff, the infographic. Whereas the poster, after many years of use, has become a very traditional form of IMRD presentation, the infographic has emerged as a highly flexible, highly visual genre both in print publications and on websites. The chapter will show how STEM communicators can use both related forms of presentation creatively and effectively.

CHAPTER TEN

CREATING POSTERS AND INFOGRAPHICS

Topics in this chapter:

I. Posters and Infographics—Using the Two-Dimensional Display Space
II. Thinking Rhetorically about Posters
III. Posters—Up-Close and Personal (or Not)
IV. Making Your Poster—Steps in the Process
V. Infographics—Thinking Print
VI. Infographics—Thinking Online
VII. Making Your Infographic—Steps in the Process

At academic conferences where research is presented, in addition to conference talks you will often find research posters. Posters are a traditional way of displaying research in a highly visual, but condensed, format. They enable text, but they feature emphatic graphics, usually charts and diagrams, and often colorfully. Posters are a great way for researchers to begin sharing their work with others in the field. This chapter explores creating effective, varied poster formats that pay attention to your **purposes** and **audiences**.

The chapter explores ways of communicating science visually beyond the traditional genre of the poster. We will examine the **infographic**, which has become increasingly popular to display research results in a condensed visual format. Whereas text is still the primary tool of the poster, the infographic emphasizes the visual, though it can still display a surprisingly large amount of text, and has the capability of linking to larger textual explanations—and other visuals.

We will also look at how social media tools are providing opportunities to share science in a visual mode. Sparking ideas for creating posters and infographics of diverse kinds is the goal of this chapter.

I. POSTERS AND INFOGRAPHICS—USING THE TWO-DIMENSIONAL DISPLAY SPACE

All the genres of STEM writing that we've studied so far in this book—IMRD articles, research reviews, STEM journalism in magazines and newspapers, and blogs—assume a multipage or multiscreen space that readers move through. Readers turn pages, scroll from screen to screen, or click on links that move them into other virtual spaces. Posters and infographics are different: they inhabit a rigid two-dimensional space—like a painting or an individual photo or drawing. Readers expect to get it all in front of them in one visual field.

This is certainly true of posters, which are physical objects akin to paintings on a gallery wall or maps and signs in a park or along a highway. The available space limits greatly what the poster can contain and convey. It forces us as writers/designers to make careful choices based on the **readers** we want to reach and the **purposes** we want to achieve.

Infographics in online environments are much less restricted, because we can manipulate the size of text and objects—and we can insert links to take the reader-viewer to other virtual places. But keep in mind that we use infographics—rather than articles—precisely because we want readers-viewers to learn something from the graphical relationships on the screen. As we get into looking at and analyzing sample infographics, we'll see how much the power of the genre depends on readers-viewers seeing those relationships as they scan the screen or the physical page.

Posters, information graphics, and other visualizations introduce some accessibility considerations we will want to keep in mind. Some people might benefit from visuals if they are more visual thinkers. People who are blind or have low vision may benefit from the inclusion of alt-text which describes what is represented in the image, or you may ensure that any text printed on the poster uses large text and images, etc. It is important to look for the most up-to-date accessibility approaches, especially as technologies change, to ensure you are considering the best way to support your audience.

MAJOR TYPES OF POSTERS

Posters as IMRD Research Reports

Posters at academic conferences typically report research using the IMRD format, as would an article for a peer-reviewed journal. But as you look through many posters, you'll note that poster makers use many variations on the IMRD format: some have no abstract, others give different names to sections of the poster. Watch for these differences and think about what naming and ordering will work best for you.

Because writers/designers want to use every square inch of the rectangular space, they treat the surface of the poster as if it were several pages of a journal, with each of three or more columns representing a single page, with the "pages" side by side.

The effective IMRD-based poster makes sure to label each section clearly—always in **bold and large-font** print. An introduction is followed by the materials and methods, then the results, and then the discussion (which might be subdivided into implications and conclusions, or something similar). A shortened references list often follows in smaller print. A brief acknowledgments section is also often included (see Chapter Five for more on IMRD article structure).

Readers of IMRD articles will expect this format and order, and they will read the poster much as they would the article.

However, because the poster is a **single space**, not an actual succession of pages, readers will tend to scan visually across the surface and be drawn to different parts of the poster—especially to attractive graphics—rather than just read the text from section to section. (A highly disciplined IMRD reader may be able to read the poster from section to section, but keep in mind that the single surface of the poster will tend to draw readers to other sections.)

Moreover, the single large surface means that graphical elements, such as photos, tables, charts, and drawings, must be arranged on the surface in a visually pleasing way. In other words, in a way that allows each graphic to stand out in its space, rather than to cluster in one section of the poster and create visual clutter.

So even if the IMRD text of the poster follows section by section in columns, the need for **visual balance** means that the space of the poster is designed more for the "viewer" than for the reader per se.

As we study the process of poster design later in the chapter (Section IV), we'll learn to apply this blend of text and graphics toward achieving visual balance.

Posters as Summaries of Research

If the poster is a summary of research, then the poster designer can choose from a wide array of possible formats and arrangements of text and graphics. More and more companies offer poster-design services both for physical poster displays and for online displays, so the researcher has many designs to choose from. We recommend scanning the possibilities to see what might work for you.

If a particular conference publishes guidelines for posters that prescribe a specific format, then the designer should follow those guidelines. One such format could be the IMRD organization that we saw for some research reviews explored in Chapter Seven. In that case, the writer-designer should keep in mind that **blend of text organization and visual balance** described just above for posters as IMRD research reports.

But if, on the other hand, no guidelines are published—or if creative freedom is encouraged—then the designer might think of the poster design space more like that for an infographic, which we'll explore in Sections V through VII below. In such a poster, the goals will be visual balance, to be sure, but even more important for the designer will be to think about

- the balance between text and graphics, and
- the poster space as primarily a visual space,

where any text is minimal and helps to support the visual impact—even though that text is still important to the readers' understanding of the visuals.

Looking at a Sample IMRD Poster

Imagine that you are at a conference and you approach this poster from a distance. Visually, what jumps out at you? Perhaps it is the university **logo** or "fish fraud" in the title. Or maybe that big red dot. Perhaps the section titles or the generous white space surrounding the colorful graphics. **What attracts you to look more closely?** Perhaps the theme *detection of fish fraud* sounds intriguing. Perhaps you like that the IMRD format is strictly followed. Perhaps your eye is drawn to the **bright colors** in the middle of the white spaces. Perhaps the large logo of the University and its Institute comforts you with the confidence that this poster reports solid research.

If, as is typical at conferences, this poster is one among many, one or more of these elements may catch your eye—and you will come closer for more detail—and perhaps to speak with the presenter standing beside the poster.

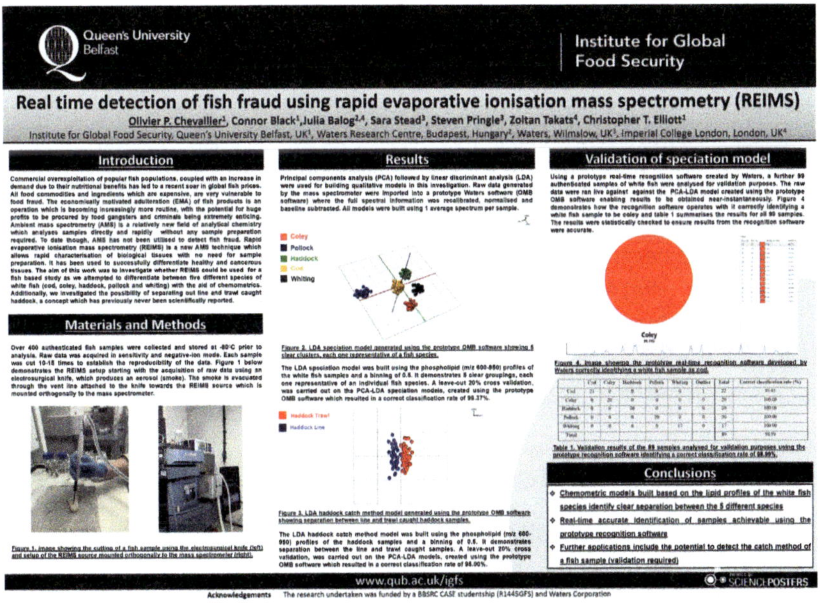

Note that from a distance of even a few feet you can't see the names of the researchers, their organizations, any detail of the figures, the photos, and the table—or any of the text—so what brings you closer to the poster are those **bolded**, large, and colorful elements.

Exercise 10A: Browse one of the many poster-design sites where you can see countless posters in a range of subject areas. Or, better yet, go to a place where many posters are on display, such as an academic conference or a student research fair at your college or university. It is also common for research departments in universities to display posters by graduate students and faculty in departmental hallways. Stand at a sufficient distance from the posters so that you can't read the text (or keep online posters small enough in size so that you can't read text). Which posters draw your attention as a viewer? Why? Note features in a notebook or tablet.

Some design factors to consider include:

- Balance of text and figures: does one seem to overwhelm the other?
- Balance of text and white (aka "negative" or blank) space: does the amount or "clutter" of text make the poster seem tedious or too

time-consuming to read? Does the lack of text make the poster seem superficial?
- Readability of the title and section headings from a distance.
- Clarity of graphic elements: does each stand out in its space on the poster? Does it seem as if you could digest the information of each figure without spending a great amount of time? For example, charts, graphs, and photos seem better able to convey information in a poster than a table with many columns and data points.
- Color: what *tones* do the colors convey? Cheerful? Serious? Dull? Interesting? Do the colors seem to enhance readability or to distract from it?

II. THINKING RHETORICALLY ABOUT POSTERS

REACHING READERS—DESIGN

Whatever draws you to a particular poster in a room full of posters is a major part of the poster's appeal to you as an audience. **A poster display is a competitive event**. As you look through posters, whether at conferences or on poster-design sites, always look for designs that speak to an aesthetic that interests you. Talk with others about their favorite ideas for poster design—and for designs they don't like. Just as different people are drawn to some artistic styles more than to others, so with posters and infographics. Learn through looking and through conversations.

REACHING READERS—LEVEL OF LANGUAGE

Now that you have come up close to an attractive poster, you can read the text and the graphics. As you read, you'll pretty quickly see if the poster writers think of you as an audience. As with IMRD articles, many posters will be heavy on insider jargon and acronyms that only fellow researchers in that area can understand. If that is you, then you'll get absorbed in the reading. If it isn't, then you may quickly lose interest and move on.

For example, in the sample poster, the title "Real Time Detection of Fish Fraud" is accessible to a broad readership, but is that what you find when you start reading the text? Might you find acronyms and jargon that make you feel outside the audience for this poster? What audience choices have the authors made? See the next section of this chapter, "Up Close and Personal," for more on ways to reach readers with posters.

What about purposes? You can't tell too much about a poster's purposes from a distance. The large-font title of the sample poster does tell you the theme of the poster in plain language that will be understandable to most

STEM people and many others—but you need to get closer to read the text and the figures. Only then can you know the purposes of the poster and the research.

As in the IMRD article, the most explicit statements about purposes should be in the **abstract (if the poster has one) and the introduction.** Note how the title and the introduction of the sample poster immediately state the purposes in regard to "real time detection of fish fraud." In a well-ordered poster, as in a well-ordered article, each section should follow from the purpose statements in the introduction. Indeed, getting quickly to the purpose statements is even more critical in the poster, because of the lack of space.

What about ethos? Remember that no one expects a poster to be as detailed and thorough as an entire article, so your credibility in a poster does not depend on those qualities. In fact, a cluttered poster in tiny fonts and with umpteen hard-to-fathom figures will jeopardize your credibility, as it shows lack of consideration for your reader.

Don't think of your poster as a *substitute* for a full article. That kind of thinking leads writers to try to cram too much information into the space. Rather, think of the poster as a more detailed and visually attention-grabbing abstract, or even as advertising for the research. You want your reader to start with the poster—and then want to learn more.

Many factors in posters can enhance your ethos—and can give readers ways to learn more if they want to. Ethos-enhancing factors include the following:

- A well-ordered design that shows the reader how to proceed through the text. Section headings are important in this regard.
- *A clear logo that shows the institutional home or sponsor of the research and researchers.
- *Listing, as in an IMRD article, the affiliations of the authors of the research (usually just beneath the title).
- A concise section of materials and methods that describes the most important elements of procedure that bear on the study (photos of instruments or a flow chart of process can be helpful here).
- A concise description of results that notes the major findings, often expressed in bullet points. (A very few easy-to-read statistics that illustrate the results are important.)
- A concise listing of conclusions (implications, recommendations, etc., labeled as such) that directly follow from the introduction and from the results.
- **A brief list of references or recommended other sources, often in smaller print.

- ***Particularly important*: if any articles by you or colleagues about your research have been published, these need to be listed in your references.
- **The URL of the project website—or a **QR code**, to copy on phones.
- **E-mail addresses of researchers (but only if you want this type of follow-up contact with readers-viewers).
- *Acknowledgments in a very brief section, often in smaller print, of funding sources or of contributors to the research not listed among the authors.

Each element marked with an * builds your credibility by showing **related work or support for your research** within the larger scientific community. Each requires very little space on the poster, but can have an impact on reader respect for your work far beyond the little room required.

The four items marked with ** give readers ways to learn more about your research **beyond the poster**. Since most readers of your poster at a conference will have digital devices, they can quickly turn their interest in your poster into finding more in-depth information about your work—especially if you have a website and if your published articles are open-access.

III. POSTERS—UP-CLOSE AND PERSONAL (OR NOT)

You and your poster are a team—at least for part of the life of the poster. At certain times during a conference, you are expected to be beside your poster so that visitors can ask you questions about your research. At other times, however, you won't be there and so the poster will be a stand-alone introduction to your research. During these times, the poster must speak for itself. And after the conference, posters are often displayed in hallways or offices of academic departments as examples of projects being done by the unit. Lastly, a poster might also be included in a website on the research, or linked to researchers' CVs. Therefore, you need to design your poster to suit all these situations.

WHEN THE POSTER MUST SPEAK FOR ITSELF

Since the poster will likely live most of its life without you to explain it, design it to speak for itself. All the advice in Section I of this chapter is meant for that purpose, so if you follow those suggestions, your poster should have a happy existence.

Do keep in mind, though, that the life of any poster will only be as long as the research reported on it is relevant. If you intend for the poster to continue its life online as part of a website on the evolving research, be prepared to **update** the poster with new text and current figures (with a note about the changes to maintain the scientific record).

Likewise, if the situation for displaying the poster changes, you may also have to **revise** the poster text to reach a different audience. For example, if the poster begins life at a specialist conference for researchers in your research area, you'll comfortably use terms and metrics accessible to that specialist audience. However, if the next venue is a more general STEM conference, or a conference in your discipline that draws researchers from many subfields, then you may need to **design and print a new poster** that will be accessible more generally. Again, don't assume that you will always be there to explain any terms that may puzzle your visitors.

WHEN YOU AND YOUR POSTER ARE A TEAM

When you are standing beside your poster at a conference (never in front of the poster—that blocks visitors' views), remember that **you are not the main focus of attention**, but an aide that visitors may call on. If you've designed the poster to speak for itself, then that should reduce your worry about having to explain what's on the poster.

Nevertheless, many visitors to poster sessions come with the intention of engaging the researcher in conversation about the project. And if viewers get interested in the research because of the poster, expect their questions. You won't be able to anticipate every question that a viewer might ask, but you can prepare for some that are likely.

For example, some viewers will glance at your poster, but they really want to hear from you. So, you may get the general request, "Tell me about your research." Rather than be annoyed that the person did not read the poster, have ready—and practiced!—a **thirty-second to one-minute introduction** to your project that covers much of what is in the introduction on the poster.

Don't repeat verbatim what's on the poster, however, but give a conversational version. This is where you can illustrate the energy and passion behind your research through word choice, tone, and non-verbal cues. Not everyone naturally has these skills, but you can learn them. Think about presenting this way as a genre in itself, which can help you structure the quick talk, and which also gives you a template to work from when the spotlight is on you. At the end of your short talk, you might look at your poster and briefly indicate the different sections that the viewer might be interested in reading. (Here's where clear section titles are important.) Then say that you'd be happy to answer questions.

Keep in mind, also, that some of your visitors may be blind or have low vision, and so will rely on you to speak in more detail about the research, because reading the poster may be difficult. While a strength of posters is their visual expressiveness, that is also a limitation. You'll need to be prepared to make up for this limitation by being a cordial, helpful, informative conversant. In terms of accessibility, you will also want to pay attention to your speaking voice, as someone with hearing loss, for example, may not be able to follow you if you're speaking softly in a noisy conference environment. Others may have sensory or auditory processing disorders that make those environments difficult, and as a speaker you can try to pay attention to see if you may need to speak louder or otherwise accommodate your audience.

You should also be prepared to answer questions about the **main sections of the IMRD format**. Expect questions about methods and materials, results, or conclusions. Since you don't know what exactly any person will ask, or what level of knowledge of the research a person might bring to the poster session, listen closely. The best preparation for these sessions is to know your research so well that you can give an informative answer to almost any question.

On the other hand, if you don't have a ready answer to a question, don't pretend that you do. Remember that these sessions are **opportunities for thinking and learning** for both presenters and viewers. It's not your job to be a know-it-all—you don't need to put that kind of pressure on yourself. Saying "that's a good question" or "we've been wondering about that, too," can be a way into a good conversation that can be helpful to your project going forward.

Aids to have with you. In addition to the sources of further information on the poster itself, you might have **brochures** or **flyers** about the project to distribute to poster viewers, plus your own **business cards** for those who may wish to contact you. Another helpful tool is a **laser pointer**, if you want to refer to sections of the poster without moving in front of it. You can also print out the text in large print for those who might need an accessible version.

IV. MAKING YOUR POSTER—STEPS IN THE PROCESS

STEP ONE

Start by "Storyboarding." Imagine potential layouts on a 24 by 36 inch (or 36 by 48) space. If you wish, use the template of IMRD poster features on p.259 or look at the example on p. 253 of this chapter to make sure you include everything important. You can use your poster design software (e.g., Microsoft PowerPoint, Adobe InDesign) to do this first-stage design, but it may be just as easy for you to draw a crude mock-up on a piece of paper in the proportions of the posterboard you'll be using.

As you draw, keep in mind the **design factors** on pp. 253–54 in Exercise 10A of this chapter. For example, in order for you to achieve balance between text and graphic figures, you'll need to consider which graphics—and how many—you can afford to use from your research so that you don't clutter your space with graphics and you leave room for text.

Likewise, plan how much space you can give to text sections, so that you don't have to use a tiny font too difficult to read (**16 to 20 point** is a typical readable size for poster text) and so you can leave some **white space** between sections and around figures. Will you use the IMRD sections? Include an abstract? How will your layout show the reader the order in which sections are meant to be read?

This process should result in a poster layout divided into blocks for the

- text sections,
- graphics,
- title and author section,
- logo,
- references, and
- acknowledgements

(or whatever elements you want to include).

Here's a first draft of an IMRD poster layout incorporating all of the features noted above:

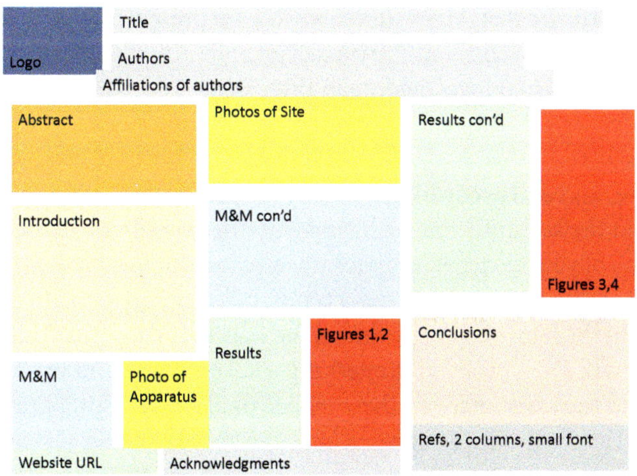

STEP TWO

Next, using your design software, begin inserting your chosen graphics into the frame. This step will show you how much room your graphics will take to be easily readable, and therefore how much room you'll have for text. Try out different sizes of each segment of the poster.

STEP THREE

Begin drafting text for each section that you think will fit into the allotted spaces. Keep in mind the principle on p. 251: the poster is not an article about your research, but a large attention-grabbing abstract or advertising for the research. Don't try to cram too much information into the allotted space.

Try drafting each section in Word using 16-point type, so you get a sense of how much room each section is likely to consume. Adjust the number of words accordingly. Remember also that you can save space by writing your sections, or some of them, as bullet points. (Note, for example, how the Conclusions on the sample poster [p. 253] are written.)

STEP FOUR

Input your drafted text sections into your frame, using your design software. Now you are ready to see what the poster might look like when completed. Be prepared to adjust sizing and word counts until you get what you think looks good. If you have studied a wide range of posters, as well as the principles in this chapter, you'll have begun to develop a good "eye" for poster design.

STEP FIVE

Get feedback on your design from (1) experienced poster makers and (2) intended readers of the poster. Most students who are presenting for the first time would benefit from some advice on design and on text from their teachers or from the principal investigators in their labs.

STEP SIX

Revise, reconfigure, and edit accordingly. Once your poster is reconfigured visually and the text revised until you are happy with the overall design *and* the quality and tone of your text, get feedback once again. But *this* time, ask your qualified reader(s) to look closely for any typos, misspellings, and other errors in the texts, in the figure captions, and in the figures themselves. This close proofreading will always turn up something you've missed, and you'll be glad that you had the assistance of others in this final phase of preparation of your careful and pleasing design.

STEP SEVEN

Practice giving your short presentation as you stand by your poster. Practice by yourself the 30-second-to-1-minute description of your research (see p. 257) that some visitors may ask for. If you're using a laser pointer, practice with that also. Then, practice with one or more persons acting as visitors to your poster, not only listening to you, but also asking you questions about your research. If more than one of you will be giving poster presentations at conferences, offer to act as a visitor for them, also.

Exercise 10B: Carry out Steps One and Three of the poster design process for a project on which you are working. (If you are comfortable with poster design software, use the tool(s) you know, but you don't need the software in order to do this exercise.)

Be sure to follow the design principles described in this chapter. In carrying out Step Three, be guided by two main principles:

- the need to conserve space—therefore, be concise
- the audiences you want to reach—therefore, adjust the amount of technical language and reliance on numbers and mathematical symbols

(**Hint**: remember that you can use the technique of audience splitting described in Chapter Five, if you want to reach both specialist and non-specialist readers.)

Also, to get a sense of how much you can write, count words in the text sections of posters you particularly like from among those you have browsed. For example, the texts and figure captions in the sample poster on fish fraud total fewer than 700 words.

If you are doing this exercise in a class or workshop, share your design and text with one or two others and ask for suggestions toward revision.

V. INFOGRAPHICS—THINKING PRINT

Infographics are even more visually focused than posters. Not only do graphic elements occupy most of the page or screen, but the visual relationships in the infographic also convey an essential idea or argument.

For example, "Warming Stripes" (Hawkins 2018) provides a simple way to communicate about the complex topic of climate change. In his blog, "Climate Lab Book" (https://www.climate-lab-book.ac.uk/), Professor Ed Hawkins posted "Warming Stripes" on May 22, 2018, along with numerous examples of how warming can be visualized. The following "Warming Stripes" graphic is titled "Annual Temperatures in Toronto from 1841–2017" and an associated text caption explains "The colour scale goes from 5.5°C (dark blue) to 11.0°C (dark red)" (Hawkins 2018):

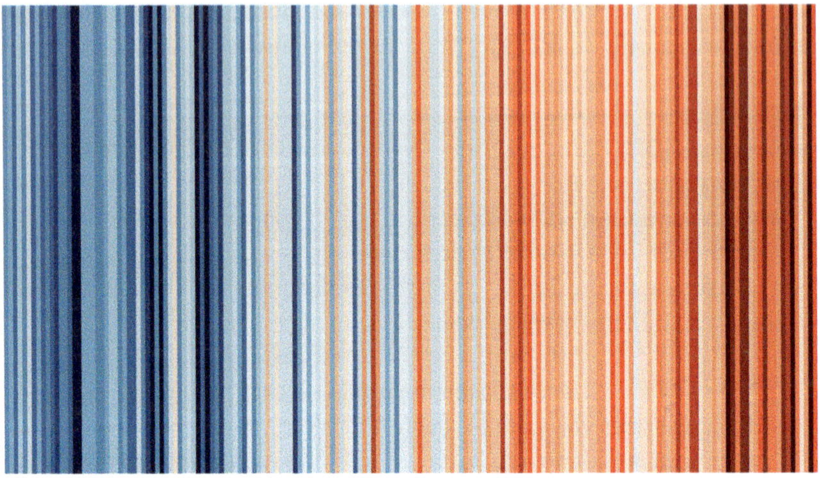

Warming Stripes, Ed Hawkins

The central idea—that the average temperatures have risen steadily over time—is conveyed visually by the graph, which also shows the temperatures year-by-year between 1841 and 2017. The years are important because they follow the Industrial Revolution through to the present day, but it is the colors in the visual that illustrate the trend so powerfully. The words provide the necessary context, but are less essential to the main message.

Infographic tools are being used in print publications today more than ever, particularly in STEM journalism. Modern designs and applications build on the two main contributions of the old infographic tools: memorable visual images and precise, vivid display of relationships, which reach many types of readers in ways that **text alone cannot achieve**. Especially in STEM-relevant journalism, infographics carry much of the weight of explanation and persuasion. Infographic tools such as timelines bring the idea of the "map" to relationships in time.

Another common tool, **color coding**, applies the ideas of measuring diversity and intensity. For example, in the map of radiation readings (Safecast 2025) below, notice how color shadings show the levels of elevated readings and the careful attention to color selection (e.g., no green indicating "safe" or similar common associations that might be misinterpreted). Indeed, we want to think carefully about design choices and the meanings they might convey to an audience. We also want to be mindful when communicating with global audiences what kind of cultural meanings might be built into our own assumptions. This is important as science becomes increasing global and as the example of Safecast illustrates (risk and hazards don't know borders and nuclear disasters have global impacts). In the case of Safecast, think about what green might indicate to audiences where the color is often associated with "go," or which might indicate safety when that's not really something that the map can convey. Careful consideration in our designs require that we understand both our own assumptions, and also understand our audience and their needs.

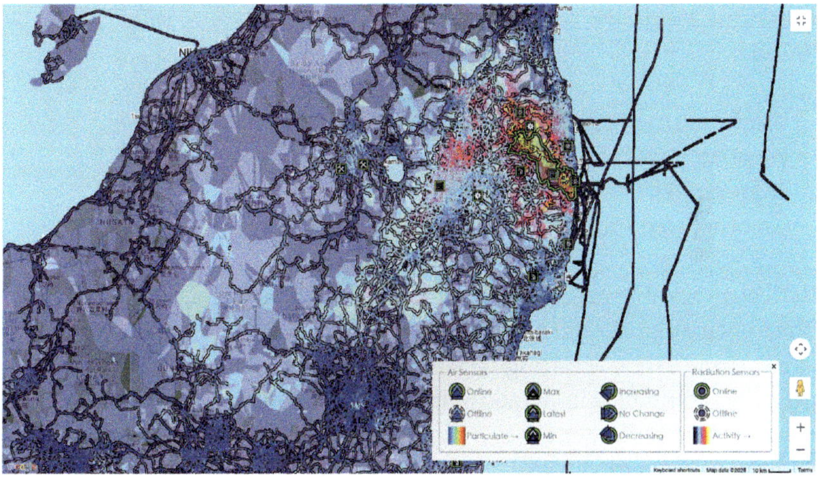

Radiation Levels, Safecast

VI. INFOGRAPHICS—THINKING ONLINE

Digital tools and online platforms allow for new ways to create and share informational graphics. The infographic below was generated with Microsoft PowerPoint and would work well in a print format. The minimal amount of text allows the photographic details to be seen, but doesn't allow further

Courtesy of Chris Thaiss

information about each plant or about any other characteristics of the site. A print publication might use the graphic as a key for individual text sections about each plant or other garden feature.

Online, however, the infographic becomes a graphic surface for **links** to a world of information about each element of the garden site. For example, note that below the titles "Euryops" and "Lupine" is the word "More." This word indicates to the reader a link to further information, which might be found

- elsewhere in the same document that houses the infographic,
- on other pages of a website created by the writers/designers of the infographic, or
- on other web sources entirely.

Each plant or other garden feature might be similarly linked to other sources. The information about each plant in the garden might be brief or it might run to many pages—and might involve further links to other sources. In any case, the infographic itself retains its integrity and visual power to illustrate relationships and other features more difficult to do through text.

Ease of changes and updates. Another benefit of the online home is that any of the linked information can be easily changed or updated, as can the text of the infographic itself. Depending on the infographic design software you use, changes to the infographic may be quick or more time-consuming, but even simpler programs allow a substantial number of adaptations.

Thinking rhetorically, note that the **infographic-plus-links** design allows you to reach many different audiences—depending on the content and style of the infographic and of the linked material. For example, the names of the plants in the sample infographic above are those familiar to home gardeners and correspond to names for plants that are sold in print and online catalogs, as well as in home and garden stores. Linked material might be tailored to the same audience.

But other links from the same plant type might be tailored to researchers studying specific varieties, while still others might be tailored to non-gardeners interested in related topics, such as landscape design. Audience splitting (discussed in Chapters Five through Seven) is made quite easy by the infographic-plus-links model.

We have discussed elsewhere how online image and video sharing platforms can facilitate scientific communications. Infographics can be found on those platforms, and we can also look at an example of how NASA created infographics by combining the traditional approaches to visualization with the benefits of video technology. An example from NASA's Scientific Visualization Studio (https://svs.gsfc.nasa.gov/) is "Annual Arctic Sea Ice Minimum Area 1979–2024, with Graph" (NASA 2024). This video shows how sea ice patterns change over time, from a vantage point in space looking down at the Arctic Sea. A graph overlays the image and as the video proceeds the pattern on the map changes and the graph's data are plotted. Below is a screen capture from the end of the video:

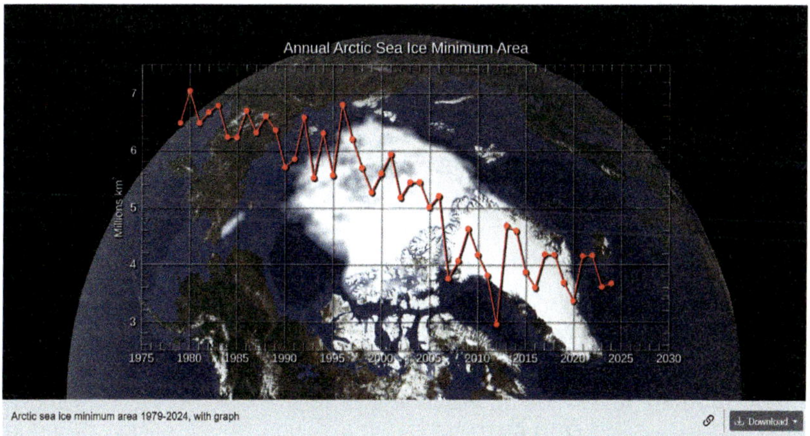

Screenshot of "Annual Arctic Sea Ice Minimum Area 1979–2024, with Graph," NASA/Goddard Space Flight Center Scientific Visualization Studio

Exercise 10C: Using the NASA website database of Images and Media (https://www.nasa.gov/nasa-brand-center/images-and-media/), explore their 3D/Tactile objects such as "3D Printed Universe" or the "A Universe of Touch and Sound" objects created for blind and low-vision individuals.

- Select an example and think about how the design process above for posters would need rethinking for these objects. Jot down some ideas about how you could begin to describe the design process for these objects. Compare your ideas with those of others.
- Then imagine different readers whom you might want to reach and how different modalities might better connect with those audiences. Identify these audiences and list types of information that might be useful to each. Again, compare your ideas with those of others.

VII. MAKING YOUR INFOGRAPHIC—STEPS IN THE PROCESS

STEP ONE

Start by imagining an image related to your own STEM research or interests that you think could make a **strong visual impression** on your intended readers. If you have not already done so, browse

- STEM journalism articles, print and online,
- STEM websites, blogs, Instagram, TikTok, etc.,
- STEM IMRD articles, and
- image files gathered by your favorite search engine according to topics you enter.

Look to these sources for images and infographics that you feel make the kind of impression you would like to make. You might use these as examples to think about what you might design.

Depending on your confidence in using infographic design software (many tools are reviewed and available online, some of them free), you can be as ambitious or as "just getting started" as you choose. Keep in mind that even something as simple and common as a **line graph, scatter graph, bar graph, flow chart, line drawing**, or **photograph** can be a powerful visual image—if it is easily readable by your intended readers (see Chapter Three for more advice on using these tools).

STEP TWO

Once you have chosen an image to make the focus of your infographic, consider—perhaps using writing-to-learn techniques (see Chapter Two)—the **purposes** you want to achieve with your image and the **readers** you want it to influence. This step will help you understand why you have chosen this focal image. It will also help you begin to imagine

- text that you want to include with your image, and
- other images that you might also want to include in the infographic, such as a timeline, a graph, or a simple table. (But avoid cluttering your infographic with too many objects, which may distract your reader.)

Always remember that the more text or other images you include in the infographic, the more you risk confusing your reader and drawing attention away from the main image.

STEP THREE

Begin to work with your infographic design software to draft the infographic. Start with the main image, then begin to add text. If you are just getting started with infographic design, be patient as you learn the software's features and quirks. You'll likely have a steep learning curve, if you are trying for effects you've not achieved before. If you have an opportunity to work with, or just get advice, from a more experienced user, take advantage of it. As you work with trying to realize your vision, don't hesitate to revise your idea, especially as you discover an effect that may be even more successful than what you imagined.

Particularly when beginning to work text into the infographic, experiment with different fonts and with different amounts of text—always keeping in mind your intended audience(s) and the need to avoid clutter.

STEP FOUR

At this point, get feedback on your draft. As with posters, the most useful feedback will come from

- more experienced infographic designers, and
- potential viewers in your intended audience.

As always in seeking feedback, ask your reader-viewer **specific questions** about aspects of the design that you are concerned about, rather than the utterly vague "What do you think?" Second, assure your qualified reader-viewer that you are indeed looking for suggestions to improve the design, not just praise for what you've done. Third, invite your readers-viewers to ask you questions, so that they can get a clearer sense of how they can give you the best advice.

STEP FIVE

Continue working on the infographic, following advice from the readers-viewers.

If the infographic will be for a print document, carefully adjust the amount of text according to its readability for your audience. In trying to cram as much text as possible into infographics, some designers use smaller and smaller fonts and text placements that make reading very difficult, especially for the visually challenged.

If you are working in an online context, this would be a good time to begin designing your linked materials and their display. As you create these materials, you'll be able to judge if any of the text on the infographic itself will need to be revised.

STEP SIX

Having now revised your print infographic, preview it by inserting it into a print environment as close as possible to how it will appear when published. For example, if it will be published in a Word document, preview it in a font size close to what your actual readers will see. This step will help you see if adjustments will be needed in the size of text fonts (and other features) on the infographic.

For online infographics, preview in a screen size that will be most common for readers. While you can't anticipate all screen variables that readers will encounter on different browsers and devices, you should do enough previewing to see if there are some formatting difficulties you can overcome by simple adjustments to your design, such as font size or spacing.

STEP SEVEN

Time for more feedback. Seeing the infographic in the context close to that in which it is likely to be published will help readers-viewers see if more needs to be changed in text or design.

Exercise 10D: Carry out Steps One and Two of the process of your infographic design. If others besides yourself will be doing this exercise, you might share good examples of infographics that you find. This process will help you consider the kind of design that you might like to achieve in your own project.

For Step Two, write informally for yourself about the purposes and audiences for the infographic you envision. Think about both visual features and text. Will you be publishing this infographic in a print medium or will you have the flexibility of online linking? Write about the implications and capabilities of the medium you choose. Be guided by the advice in this chapter.

If you will be doing this exercise with others, share ideas from your writing with one another and ask one another questions that can help each of you refine your vision.

CONCLUSION

This chapter has offered ideas for your design of posters and infographics: one a traditional medium for the presentation of STEM research in a concise format, the other a newer and more versatile medium meant to make a strong visual impression through an increasingly wide range of technologies.

What this chapter has not tried to do is to delve into the many hybrids of poster and infographic design, such as

- highly visual print posters (infographics, really) that STEM communities use to advertise events, research opportunities, and exciting new degrees and companies, or
- posters in online environments that present IMRD formats, but that have the capabilities of linking to related sites and references.

As technologies evolve, we can expect to see more merging of these concepts, as well as new designs that bring the visual—and auditory—arts into the service of the communication of science, and into the scientific process itself. We talked about some of these in the chapter on multimodal forms and we are seeing video become even more common in the communication of science.

In the next chapter, "Creating Oral-Visual Presentations," we explore another traditional STEM format that is ubiquitous in academia and the workplace, and that more and more offers opportunities to take advantage of new technologies. At the same time, this focus will let us think about the role of the human presenter—as teacher, performer, resource, team member, and personality.

CHAPTER ELEVEN

CREATING ORAL-VISUAL PRESENTATIONS

Topics in this chapter:

I. Presentations as Unmatched Opportunities
II. "Presence" and "Being Present" in a Presentation: Two Scenarios
III. Making Your Audience Your Ally
IV. The Visual in Oral-Visual: Striving for Balance
V. Achieving Success through Preparation

Public speaking is a common fear for many people, but it is a critical skill for many STEM students and professionals. Whether to make presentations of research to small groups of colleagues, to audiences at conferences, to fellow students and professors in their classes, or in more intimate, but still high pressure, occasions such as job interviews, reports to the boss, or pitches to potential clients or grantors, STEM students and specialists require skills to become oral presenters. While public speaking can be frightening, you can gain skills that will allow you to become effective and perhaps even comfortable in these genres of science communication.

Chapter Ten provided a segue to this chapter when we considered the vital role that you play in poster presentations when you speak to visitors about your research while at a conference (pp. 257–58). That occasion provides a good example of what a researcher must be prepared for in anticipating a speaking role, though we did not, in that chapter, go into the anxiety that many might feel in this somewhat ambiguous presentational situation.

This chapter will give advice on ways to overcome the various fears that may arise in STEM-relevant presentation situations. Moreover, it will offer ways to make oral-visual presentations of your research a highly useful contribution to your STEM work—and a vital opportunity for communication *and* influence not equaled by other media.

I. PRESENTATIONS AS UNMATCHED OPPORTUNITIES

Most of this book so far has focused on how we communicate through writing, as defined multimodally and across digital media. You imagine yourself in front of a computer screen or with a tablet in hand, and that's how you communicate. You take some time to craft your words and pictures, get feedback, revise, and then you send them out into your online community, or to your email lists, or via your blog, website, written report, or article. What all these media have in common are

- time to prepare your message,
- time to get feedback and revise, and
- your ability to hide behind the device and present a *mediated self.*

Presenting *in person* takes away those screens of time and place, and therefore *exposes* you—including your voice, your facial expressions, your body language, and all aspects of your physical appearance, plus how you react emotionally to what other people say and ask and how they react to you. So, no wonder in-person communication can be scary—especially if you spend most of your communicative life behind those screens.

But, as we explored in Chapter Two, communicating through writing and its associated media creates its own fears. The greatest fear for you (and us) as writers is that, no matter how much time we spend preparing our messages and weighing our sense of audiences, the mediated self we craft so carefully might get it wrong. Rhetorically, we'll fall flat.

And, equally scary, once that mediated self is out there, it's hard to know exactly how people are reacting to the mediated you. Sure, you can count "likes" and "followers" and "hits." You can sit waiting for an editor's response to your article or a funder's response to your proposal. But all these measures not only take time, they also give the responder (or the social media "friend") time to make a *mediated response*—one that presents a partial message that's often hard to understand or trust.

How often have you wished that you could talk to an editor or a reader, to get a fuller sense of what their response—once you get it—means? How often do you look forward to talking, person-to-person, with friends or colleagues with whom you can share your ideas, who can give you feedback you trust, and who can ask questions that show their interest (not just a "like") and their desire to learn more? How often do you need that opportunity to show others your feelings, expressions, and reactions in real time—and to get their reactions in turn?

THE SPEAKING-WRITING-LISTENING-READING CONTINUUM

Throughout this book, the process of writing that involves oral communication in the form of **feedback** from qualified and trusted peers (as well as from instructors and other experts, when possible) has been central to how we think about our communications. This oral interaction is not meant as an add on. It manifests what has always been a vital connection among the language modes of writing, speaking, signing, reading, listening—as well as watching and sharing facial expressions and body language. Millennia ago, writing as a technology grew out of oral communication, and it remains an extension of the **languaging process**—but not a replacement for it.

Presentations are unmatched opportunities for communicating your STEM research. Even today, with various forms of online communication including video, oral in-person presentations hold an essential place. Even if the peer-reviewed article is the gold standard in STEM communication, writing and reading are always part of a **multimodal continuum** with the *many* oral-visual interactions in the STEM research process:

- conversations in the lab or classroom
- phone calls or video calls among colleagues or with fellow students
- spoken responses to proposal drafts among colleagues
- impromptu presentations of research ideas to supervisors or potential team members
- formal presentations at team meetings, or at conferences, or to clients
- all the informal and structured feedback to work in progress

In all these venues and situations, writing is no substitute for the oral-visual communication we need to advance knowledge. These media work together. And our integration of oral communication continues to transform as our media and working conditions change.

Since the COVID-19 pandemic transformed workplaces, we have all had to think a lot about the different methods we communicate in oral-visual ways both online and in person. While in-person oral communications can be challenging for all the reasons noted, we also know that it's necessary, and we need to prepare ourselves to do it without too much anxiety.

Exercise 11A: List the ways in which oral communication is a part of your work as a STEM researcher and practitioner. (Use the bulleted list just above to help in your brainstorming.) If you are doing this exercise with others, compare your lists. Note the role of technology and think about how it might change how you communicate differently than in person.

For several (or all) of the items on your list, briefly describe in writing

- how often you engage in this activity,
- what skills and knowledge you bring to this activity,
- why it is important to you,
- what you seek to learn from it,
- how it contributes to your growth as a researcher, student, professional, etc.
- any anxieties that you need to overcome to engage in it successfully, and
- how it helps you as a writer.

ANOTHER CONTINUUM: FROM THE INFORMAL TO THE FORMAL TO THE INFORMAL

In considering the formal occasions in which you must speak—at conferences or in classes or before clients and supervisors—it may help you face the anxieties you feel about the formal talk if you think about the movement from the informal to the formal and back again. Of course some people, such as some neurodiverse individuals, might prefer formal scripted talk to informal, unstructured talk. Everyone's experience is a little different, and what's important is figuring out some strategies that work for you rather than adopting any specific approach. The goal is to be able to speak confidently as an expert about your subject.

The most successful formal presenters are those who can bring to the formal presentation something of the **informal—relaxed and conversational**. For example, if you watch a TED talk (https://www.ted.com/talks) or really pay attention to a good lecturer in a class or at a conference, you'll see that the speaker comes across as relaxed and is able to make you feel as if they're speaking to you personally, not as if you're a stranger they need to be afraid of or nervous in front of. We see this trend in social media space, too, such as TikTok, where science communicators use more formal and informal styles to discuss complex issues. For example, they might have a script, but part of the delivery is to present the material in a manner that feels engaging and personal. Even in the formal, in-person environment, it is as if the speaker were talking with a small group of people who share the speaker's interests and who can learn from one another.

And, in fact, successful formal presenters almost always talk on subjects that they have already talked about *informally* with friends or colleagues, often frequently. So, they have learned and practiced, often unconsciously, the language and gestures that seem to work well in the more casual setting. A good analogy is with musical performers, none of whom began as performers in huge concerts, but who played or sang many times in and before small groups before they ever imagined performing in a concert hall or arena. Moreover, the good speaker, like the musician, doesn't think of the formal presentation as an end, or peak, in communicating with an audience. Instead, and perhaps even in the midst of the lecture or performance, the presenter is already learning from the formal event and from the reactions of the audience: learning ideas, thought connections, and language that she or he can work on

...back in the studio,
...or the lab,
...or before the computer,
...or over coffee among colleagues.

The continuum of informal > formal > informal reinforces the continuum of writing-speaking-reading-listening-watching-sharing to help the researcher develop ideas and methods. All the oral-visual interactions, including presentations, provide unmatched and essential contributions to the growth of scientists and of professionals in all disciplines.

HOW CAN THESE CONTINUUMS HELP IN YOUR PLANNING FOR PRESENTATIONS?

First, regarding any anxiety you may have about formal presentations, think of the formal presentation as just another opportunity to share your views and knowledge with interested listeners and to learn from the interaction. Experienced speakers tend to think of their audiences positively, not as enemies or as stern critics, but as people who want to learn and engage the ideas, just as occurs in informal discussions about research. When we and our colleagues help to prepare graduate students for their conference presentations, we offer opportunities for the students to give their oral-visual talks to us and to fellow students a week or more before the conference. Most students take advantage of these constructive small-group opportunities, which they say are valuable to their success. They say that they learn from the comments and questions of the attendees in our pre-sessions—and they come to realize that the small group of listeners is not much different in their attitudes and questions from the larger audience at the conference.

Second, understanding the two continuums helps you realize that all presentations, informal and formal, are steps in your growth and two-way opportunities for learning. In planning for presentations, we tend to overemphasize the one-way dispensing of knowledge, as if we are supposed to be technicians giving inoculations of facts, figures, and opinions. We tend to put all our effort into what we'll say and what we'll display for people to see. We *underemphasize* the audience as participants in and contributors to the discussion. Our tendency as planners is to try to cram as many words and slides into our presentation as possible, and in so doing we fail to leave room for the audience to think about, digest, ask about, and respond to all that we've said and shown. As a result, presenters tend to drown their audiences in "information," speak way too fast, click from slide to slide to slide, and go over their time limits—despite the pleading efforts of the conference planners to build in time for actual discussion.

Later in the chapter (see Section III, "Making Your Audience Your Ally"), we'll offer several techniques for **inviting audience participation** and therefore **increasing their attention** to what you say and display. These techniques enhance the presentation as a **mutual learning** opportunity.

Exercise 11B: Watch a few presentations in a field relevant to your research that have been recorded and shared online. These should be recordings of actual talks given to live audiences (such as TED talks or keynote talks given at a research conference). Better yet, if you have the opportunity, take part in live presentations at a conference, or lectures in a college or university.

Observe the presenters and presentations closely. In particular, for each presentation, pay close attention to how the presenter is actually "present" in the place and the moment. For example,

- How does the presenter show awareness—and perhaps appreciation—of this audience?
- Does the presenter show awareness and appreciation of the place? How?
- Is the audience invited to participate in any way? How?
- Are the content and the pacing of the talk and of the visuals adjusted to allow the audience time to think about and digest information before the presenter moves ahead? If so, how?

- Does the presenter invite questions and comments from the participants? How are these invitations phrased? How does the presenter show clear interest in the engagement of the audience, rather than a pro forma "are there any questions?" at the end of the talk?
- How does the presenter listen to and respond to questions?

Overall, does the presentation seem to you an authentic learning opportunity for both the audience and the presenter? How would you improve it?

II. "PRESENCE" AND "BEING PRESENT" IN A PRESENTATION: TWO SCENARIOS

We've all attended lectures that come across as "canned"—the speaker has prepared everything well beforehand, and proceeds to read the talk with only occasional eye contact with the audience. Meanwhile, the speaker moves from slide to slide of an accompanying slide deck—and maybe throws in a short video or two—but you feel that this is the same talk the person gave elsewhere, and you wonder if the speaker is at all interested in what you may have to say about the subject, or even cares if you understand what has been said and shown, because the talk has been delivered too hastily. The talk ends and the speaker asks, "Are there any questions?" If any questions are forthcoming, the answers also seem canned to you, and the speaker ends the event by leaving the room or hall—and that's that.

In contrast, we've also all attended lectures where the speaker comes across as genuinely interested in *this* audience in *this* setting, where the speaker has done much preparation of words and visuals beforehand—but is also willing to go "off-script" in response to a question, and paces the talk and the visuals so that the audience has time to think about and digest ideas and evidence before the speaker moves on. The speaker also takes care to make **eye contact** with different parts of the audience during the talk. If you're not comfortable with eye contract, a common strategy is to look around the room at people's foreheads. Some speakers may even *begin* the talk by asking the audience to think about certain questions that they will want the audience to respond to later. The speaker may invite questions from listeners *during* the talk, not just at the end. Moreover, when questions are asked by audience members, the speaker will appear to listen closely and may ask for a clarification so as to better understand the question and give a better answer. Indeed, much of the speaker's work in the presentation will come in response to questions or comments, and the speaker may use the questions to go back to certain slides or ideas—or share some original thinking not pre-programmed into the talk. You come away from this talk still discussing it with others and energized by these interactions.

The two scenarios described above illustrate different purposes for what are called "presentations," but only the second tries to take advantage of the *presence* of many minds and personalities in the same place at the same time.

The first scenario—the "canned" talk—strives for transmission of information not much different from what a viewer might get from a slide deck with a voiceover. It's also similar to a short version of a research article—or a summary of research—with a number of graphics. For such a "presentation," there is no need for a large group of people to come together, nor for a speaker to address them. Technologically and rhetorically, it would be much more efficient and inexpensive to post the slide-deck-with-voiceover (or the research summary) on a website for each viewer to watch or read on their own schedule. Although one-way learning may occur in this viewing or reading, there is no opportunity for the speaker to learn from the readers-viewers and no opportunity for mutual learning among the readers-viewers.

Twenty-first-century technology allows educational planners and conference directors and others to explore new ways to facilitate in-person events and create **hybrid opportunities** for sharing in knowledge production. For example, colleges and universities are adopting versions of what is called the "flipped classroom": that is, in contrast with the traditional model of using class time to deliver lecture content. The new model "flips" the classroom setting, allowing lecture content to be pre-recorded with a slide deck so that students can review the materials prior to in-class meetings. Then, when students and instructor meet in the classroom or lecture hall, the time is used for small-group discussions, Q&A, or work on team research projects. In other words, the flipped classroom takes advantage of the presence of many minds and personalities together, so that **mutual learning** can occur.

Similarly, conference planners are redesigning schedules to make room for diverse, interactive, mutual learning opportunities. Less time is allotted for traditional talks, more and more for

- poster sessions,
- workshops and "think tanks" on specified projects,
- roundtables for Q&A,
- special-interest group meetings, and
- mentoring sessions for those with new ideas—where experienced researchers listen to ideas pitched by newer scholars and then comment on those ideas.

Within "traditional" presentation sessions themselves, models are changing, too. Proposed panels by pre-arranged teams are replacing individual talks, and—more important for our theme here—proposals for talks are being judged by review panels on how the proposer (or proposing team) will blend (1) transmission of information with (2) audience participation during the session.

Exercise 11C: Study the calls for proposals for several conferences related to your field of research. Observe both (1) the descriptions of the types of sessions for which proposals are requested and (2) the requirements for proposals for each type of session.

As you observe, note the following for each type of session:

- How are the purposes and goals of each type of session described?
- How much emphasis is placed on participation by the attendees at each session?
- How much emphasis is placed on the content of talks given by individual presenters?
- How much time is allotted for each type of session, and are there guidelines for the proportion of time for individual presentation vs. attendee participation?

What does this study of the conference calls for proposals tell you about how you should envision a session you might propose? What will be happening and should happen? What does your study tell you about how you need to allocate your planning time between

- the design of your talk, including creation of visuals, and
- the design of activities for the attendees?

III. MAKING YOUR AUDIENCE YOUR ALLY

So far in the chapter, we've moved from a fear-inducing opening picture—an inexperienced presenter fretting about a strange and possibly hostile audience sitting quietly in judgment—to a much more positive scenario: a more relaxed session planner thinking about ways to make a presentation a *mutual* learning opportunity. If you've done Exercises 11B and 11C, you are already getting ideas for ways to be truly present in your presentation, so that your time with the audience can be productive for all concerned. Here are some suggestions for this process, which we call

SOME WAYS TO MAKE YOUR AUDIENCE YOUR ALLY

1. **Keep eye contact with your audience.** This is part of bringing informality to the presentation and showing the attendees that you are interested in their input and their learning. Achieving eye contact requires that you not read your prepared remarks line by line, but know the sequence of ideas well enough that you need only occasionally refer to notes or to lines on a slide deck. Eye contact also means that you try to look at folks in different parts of the room. Remember, if eye contact does not come naturally to you, that's okay and you can look at people's foreheads or around the room otherwise to engage the audience. For most speakers, eye contact in the room may take some conscious practice, because if we're nervous we tend to focus on a few audience members who are smiling at us or nodding their heads at our wise remarks. Making eye contact with different folks around a room maybe be easier if the session is interactive, which means that you'll be hearing questions or comments from a number of participants across the space.

   When we think of communicating online in video-based social media, we can think of the ways that mediated space also demands us to attend to "eye contact" or similar engagement. When we're speaking, for example, we are often mindful of the placement of the camera, especially if we have multiple displays and may be reading off one while the camera is mounted within another. Although there are certainly differences, some of the rhetorical purpose here is similar, which is to engage our audience at an individual level.

2. **Acknowledge the audience at the beginning of your talk, in particular your interest in hearing from them later in the presentation.** If you are part of a presenting team, make sure the first speaker thanks the audience for coming and for taking part in the session. Assure them of your interest in hearing their views and questions later on—this assurance will engage them in your talk and give them a **role** to be mindful of.

3. **Make sure the audience knows the plan for their participation—and then follow it.** If you've done Exercise 11C, you'll know that different types of sessions at conferences require different forms of participant

involvement. Designated workshops differ from roundtables or mentoring sessions, and these differ from panels or individual presentations. Whatever your designated format, be sure to go over it with the audience at the beginning—and stick to the schedule. Losing track of the time and losing track of the purposes of the session will frustrate your audience. A good tactic is to show a timed schedule on a slide at the beginning, spend a minute or two going over the schedule—and even ask the audience (humorously) to keep you aware of the time if you just happen to lose track. Of course, if you are part of a presenting team, each person needs to act as a timekeeper for others. Or, if your session has a moderator, that's the moderator's role.

We might also think of how audiences are asked to participate in **online environments**. For example, if we are giving a live lecture, there may be ways to ask questions online or maybe even have someone in the room read online questions. If the lecture is entirely online, it is often easy to allow people to put up their virtual hand and ask a question. Since the COVID-19 pandemic, many workshops have been held online, and some conferences now have hybrid forms where presenters may either present online or in person. In all cases, keeping time is important. Being mindful of time and timing, too, applies to prerecorded video content. Although some platforms allow for longer format videos, we need to be mindful of our demands on the audiences, and keep our messages on point and, ideally, punchy in order to keep online audiences watching. With respect to engagement, we might also let our audience know how they can engage with our content and with us as presenters. Are you open to comments or direct messages (DMs)? Carefully consider your audience, its size, and what kind of engagement you hope to have.

4. **In planning for an interactive session, be sure to spend sufficient time working out the process of the interaction and its steps.** Don't underestimate the amount of time an interaction will take up. If you are not accustomed to conducting small-group discussions, writing exercises, or team brainstorming (or other workshop-type interactions), make sure you consult with colleagues who are experienced in these methods (or have such colleagues on your presentation team). For example, a 90-minute workshop session should have *at least* an hour for activities, including groups or individuals reporting on what they've learned. Your time for talking about your research may be only 15 minutes or so.

Similarly, a roundtable that includes six or seven individuals giving short talks, to be followed by give-and-take among the speakers and the audience, will need to have speaking times rigidly enforced. Five minutes *means* five minutes, and the moderator will have to be the "cruel" enforcer. Remember that the purpose of these sessions is the audience participation, not the speaker's remarks, which are only meant to get the discussion going. Online, this kind of workshop can be effectively conducted with electronic "breakout rooms," but then you must ensure that someone is in charge to wrap up each room before you regroup. While technology allows us to force the breakout sessions to rejoin the main group, that is not an effective way to facilitate a workshop. Rather, you should be able to have the breakout sessions conclude and determine how the group wishes to rejoin and participate with the larger group based on the conversation they've just had.

5. **Consider asking "framing" questions to the audience at the start of the session.** The IMRD article structure (Chapters Five and Six) always puts front and center the purposes of the inquiry, usually in the abstract and again in the introduction. If your presentation is a report on your research, a good technique for engaging your audience is to **frame the purposes of your research as questions** that you can put on an early slide and present to the audience. You might title the slide something like "questions guiding this research" or "what we are seeking to answer." Using this technique will help engage the audience in the quest for knowledge that the research represents. Each discipline has somewhat different norms for presentations and their structure, and so it's helpful to attend presentations in your own area of research to better understand these norms. Normally, however, there will be some way that this framing work is accomplished early on in a presentation. This can be an effective hook online, too. The questions may be rhetorical in that your audience, if the presentation is pre-recorded, cannot answer. However, this can be helpful in getting the audience engaged and thinking about the problem you want to talk about.

6. **Always answer questions and comments from the audience respectfully and with interest.** A respectful, interested attitude with an audience should go without saying, but we have seen some inexperienced (and even experienced) presenters over the years react to questions and comments in negative ways, both in their words and in their body language or facial expressions. Your words, facial expressions, and body language should always show that you appreciate the question or

comment and are thinking about a helpful reply. Every once in a while, the question or comment itself may be phrased negatively, either as a challenge or even with some contempt; however, **most questions or comments are made respectfully and with a sincere desire to learn, and the speaker should respond with respect and with a desire to learn**—even to those questions and comments that are not stated in the spirit of a mutual quest for knowledge.

If you're really having a difficult time with someone, it can be effective to redirect the conversation. For example, you might say something along the lines of, "Well, that's such an interesting and complex point. I can't address all of it in the time we have, but perhaps we could chat following this session. Thanks so much," and then politely but firmly move on to the next question.

One reason why it's crucial to leave enough time for questions and comments is so that your responses can be thoughtful, so that you'll have time to go into a bit of detail in a response, or to ask the questioner or commenter to explain a bit more. If you can see that time is running short and you want to hear more questions, always extend to questioners the opportunity to write to you after the presentation, or, if possible, to talk with you briefly after the session.

7. **Plan your talk rhetorically—remember that people are listening to you in real time—they can't go back to re-listen before the talk moves forward**. An important part of making your audience your ally is being sensitive to the audience as listeners. Sure, today it's common for conferences to ask their presenters to send their slide deck to a database, so that viewers can see the visuals later. It's also becoming more common for talks to be recorded and then uploaded to online platforms. It's been a long tradition for some conferences to publish "proceedings" volumes: collections of articles based on selected conference talks. But keep in mind that for a real-time presentation itself to be useful to the audience, speakers must realize that the audience can only process so much information, and that they need time to **digest** and **think about** what is being said. It is also helpful to think about the accessibility of your talk before you present it.

If you're presenting online to a live audience, there are tools that can help you, including automated transcription and/or captioning. Artificial intelligence (AI) tools are improving how this is done, but it's always helpful to be aware of its limitations. You may also think about opportunities for in-person events to have someone to sign alongside the presenter, particularly if you're a conference planner. Or perhaps use tools to create live transcriptions that are made available to participants. You can also use older technology to effectively accommodate for different audiences, including printing out your presentation script in large font, etc.

A FEW DOS AND DON'TS

- **Limit what you say.** In Chapter Ten, we stressed that a poster can only report on a fraction of what an article would cover. Use the same thinking about a talk—say only a small fraction of what you would write in an article. Think closely about how much your audience is likely to assimilate—and how much they can remember at a time.
- **Don't talk continuously—give listeners time to assimilate and read slides.** Frequent pauses help.
- **Organize ideas clearly.** Visuals can help with this (see Section IV for more on the visual component). As you talk, make sure to **emphasize main points and repeat them as needed.** For example, if you know you want to make three or four main points during your talk, announce them up front to your listeners—much as you would in the abstract of an IMRD article. Then, if you take each point in turn for more explanation, the audience will still know where you're headed. Similarly, if you use the IMRD format to structure your talk, the audience will know what to expect as you proceed.
- **Don't get bogged down in numbers.** Limit the numbers you cite in a talk. Keep to the very few stats and trends you need to show to make your main points convincing, and reinforce them visually. Refer listeners to other sources for greater detail.
- **Don't get bogged down in descriptive details.** See the point just above about numbers.
- **Tailor your technical language to your audience.** If you will be addressing non-specialists, review Chapter Eight (about writing for non-specialist audiences). Never use a technical shorthand (such as acronyms) as a way to save time. Your audience's ability to keep track of your verbal flow and to understand what you are saying takes precedence over your saying all that you think you need to say. If possible, try out your talk beforehand with listeners from your intended audience, or listeners who know that audience well.

Exercise 11D: Brainstorm in writing a plan for a presentation that you (or a team you are part of) might give to a specific audience. Consider the following:

- How much time will you have for this presentation?
- What are the main points that you need to have the audience understand during this presentation?
- If the audience remembers little from your talk, what is the most important thing they should remember?
- For this audience, what characteristics of their interests and levels of knowledge do you need to keep in mind?

Review the chapter thus far closely. In your plan, build in things you will say and do, including conducting activities for the audience, that will enhance their engagement and learning. Block out time accordingly (as well as you can at this preliminary point).

If you are doing this exercise with others, compare your ideas.

IV. THE VISUAL IN ORAL-VISUAL: STRIVING FOR BALANCE

In Chapter Ten, we thought about the balance between text and graphics in posters and infographics. In thinking about speech and visuals in oral-visual presentations, balance is also important.

But here the issues are different, too. In posters and infographics, both the text and graphics are visual. We see the text and we see the graphics: our eye moves back and forth from one to the other. We also have time to move back and forth. We can choose where to look and then shift our vision. In the oral-visual presentation, the speech is usually being delivered simultaneously with the visuals: the words are heard and the visuals are seen, and we often have to process both simultaneously.

Moreover, most of the words are spoken and then gone—poof!—and if we've not paid attention, we won't remember them. So, if we're looking at a beautiful photo on the screen, we may miss what the speaker is saying—and thus perhaps the point of the photo!

Now, good speakers try to deal with this problem in several ways:

- **They might use the screen to show part of the text of the speech**. We see this all the time in talks, when the presenter puts a key sentence on the screen; for example,

  Climate change is happening faster than many experts have predicted.

The speaker shows the sentence and reads it to the audience, so the visual is part of the talk. Then the speaker talks more about *that* sentence, while *that* sentence remains on the screen. No conflict between what is heard and what is seen. This is a good tactic, if the speaker wants the audience to remember that key sentence, which is usually the case. We also see this strategy used in online presentations where text is overlaid with what the speaker is saying. Key words can be included in visually memorable ways as the speaker provides additional information.

- **Or the speaker might put more of the text on the screen and read all of it to the audience!** (You've seen this, right?) But *be careful.* While there is no conflict between what is heard and what is seen, the amount of text on the screen can get really long, and the visual can get really dull and boring. The listener just looks at the screen, and the speaker is tied to the visuals. No room to adlib and no one pays attention to the speaker. The viewer thinks, "Just send out the link to the slide deck. I'm outta here!" When reading longer texts in online videos, you too should be selective. This can be an important strategy when, for example, you're quoting someone on an idea you want to discuss. Again, however, you need to be mindful of your audience, of the media, and of the genre conventions. As we have suggested watching in-person presentations, it's helpful to find some exemplars for online communications that are current (genre change happens fast online!), well received (shared, liked, reposted, etc.), and, of course, informative.
- **Or the speaker uses the screen to show a captivating photo, drawing, graph, or map**—a visual that captures the viewer's attention and, equally important, that can't be replicated in words alone. You can use this strategy online as well, and there you might explore a range of visuals that include emojis and other relevant but niche strategies for a given platform, time, and audience.

Courtesy of Chris Thaiss

The speaker may then talk about the graphic, or give the viewer a short time to look at it before talking about it. As long as the speaker talks about the visual, the viewer has time to shift attention back and forth from the visual to the relevant spoken words. This is a fine tactic that uses the power of both the words and the visual.

A series of visuals may be used in this way, and the tactic will remain balanced and successful as long as

- the visuals are easy to see for *all* the audience,
- the speaker allows the viewers a short time to study each visual before speaking, and
- the speaker talks only about the visuals.

Again, the goal in blending visuals and spoken text in this way is to *reinforce the power* of each to engage the audience and convey a unified point. The audience needs to feel that coherence, or they will again be torn between listening and watching.

A FEW MORE DOS AND DON'TS FOR ACHIEVING BALANCE AND COHERENCE BETWEEN VISUALS AND SPOKEN TEXT

- **Avoid complicated visuals** that are hard for *all* the audience to see clearly and that will require a long explanation from the speaker. For example, tables with many data points are usually best avoided, unless the speaker has visually highlighted a few key points that convey an important result. Similarly, showing photos of complex apparatus with many small details that need explanation are best avoided, unless the structure and functions of the apparatus are the main focus of the presentation—and unless visuals of discrete portions of the apparatus are also shown for easier viewing.
- A **laser pointer** can be a useful tool for **focusing** viewer attention on details of a photo, drawing, graph, or map, or on specific lines of text on the screen. Being guided by the pointer will allow viewers to zoom in on important details and will reinforce what the speaker is saying about the highlighted part of the visual. Online, you can point to certain visuals or markup visuals to indicate the important features.
- While it's important for a speaker to maintain eye contact with the audience, **feel free to look occasionally at the screen** as part of helping the viewers to focus on a part of a slide. When the speaker turns from the screen back to the audience, it signals the viewers-listeners to pay attention to the speaker. Some of the sharpest moments in presentations

occur when a speaker points—either with a finger or a laser pointer—to a specific detail of a visual and then turns back to the audience to make a point about what has been identified.

- Along with eye contact, some **gestures and movement** by a speaker will keep audiences engaged during a talk. Watching talks by successful speakers will help you learn how you can use body language to establish rapport with your audience. If you must stand behind a podium or lectern, maintaining eye contact and using some relaxed gestures are especially important, so that audiences know that you are thinking of them and not just about what you'll say next.
- You can also use gestures to **indicate emotional reactions**—such as surprise—or confusions, etc., and these can be effective in online short clips. If you're a hand talker, it can be useful to think about how that relates to the way you're thinking about ideas and how to incorporate this co-speech strategy into your presentation style.
- In practicing for your presentation, build in time for **allowing viewers to look at the visuals while you are silent**. If you are using many slides, that wait time can add two minutes or more to your talk, so you'll need to accommodate it by decreasing content accordingly.
- Recent versions of presentation platforms give presenters so many design choices that you may be tempted to try out many of them. Experiment all you want, but remember that whatever group of effects you eventually choose should always help your viewers-listeners (1) **see clearly**, (2) **focus attention**, (3) **avoid distraction**, and (4) **reinforce the order and main points** of the talk. (For example, one of us likes to use colors and photos in slides, but tries to keep the color palette consistent to ensure the photos support the text.)
- If you use a **video** in your in-person presentation, do so only if it adds something that you feel is essential to your purposes for the talk. Make sure that you rigorously edit the clip so that (1) it only **supports your focus** in the talk and (2) **doesn't rob you** of time better used to explain your ideas in the talk. Rigorously cutting videos can be challenging—sometimes because the quality of the video is high and the presenter fears that the audience may like it better than the speaker's own presentation! But respect your own work—and be ruthless in cutting. Also, practice showing the video clip in the venue of the talk beforehand. We've seen too many presenters fumble with nonworking videos because they didn't practice with the equipment in the venue beforehand.

Exercise 11E: Review several of the presentations you chose for Exercise 11B. This time, pay particular attention to the visuals—including text and graphics—that support the talk. Consider the following:

- How do the visuals support the talk? Do any appear redundant, or unexplained, or distracting?
- Do you have time as a viewer to look at the visuals before they disappear?
- Which visuals seem to you particularly effective? Why?
- How might you improve the use of visuals in this presentation?

What do you learn about balancing visuals and talk from this exercise?

V. ACHIEVING SUCCESS THROUGH PREPARATION

You'll prepare for a presentation pretty much as you would prepare for other writing and design projects that we've covered in the book so far. The additions here, of course, are that you'll also prepare to deliver your project through speaking and by designing activities for your audience. So, preparation entails three closely related aspects, each of which has bearing on the other two.

Because of these three related aspects, I recommend beginning your preparation by considering the rhetorical "big picture." What this means is that you'll start by considering the **purposes** for your presentation, the **audience** you'll be presenting to and working with, and any **constraints** brought about by the setting and the event.

CONSIDER CONSTRAINTS OF THE SETTING AND THE EVENT

Begin by considering the constraints. For example, if you'll be giving a talk with slides at a conference, you'll need to work within the conference's rules for presenters: how much time will you have for speaking and how much for audience interaction? What other rules are found in the instructions for presenters? How many other speakers will you be sharing time with in the session? Will there be a moderator to introduce you and keep time for each of the presenters? Also, what presentation equipment options and capabilities will there be? How large will the room be, so how many people should you plan for? Will you be speaking from a raised platform and will you be wearing a microphone? Or are you going to be in a small conference room and sitting with others in a circle or around a table? Knowing these likely constraints will help you envision the setting and imagine what you can and can't do during your allotted presentation time.

Successful presenters always learn as much as they can about constraints either from conference organizers or from presenters experienced in this venue and those like it. Experienced presenters will know more questions to ask about constraints than inexperienced presenters can even imagine. That's why our sessions with graduate students preparing conference presentations for the first time are so valuable for them.

In contrast, let's say that you will be giving a presentation to a potential client or employer. The same process applies—but most of the answers to your questions will differ from the information you received about the research conference. Again, go into presentation planning with as much information about constraints and options as you can gather from experienced presenters.

CONSIDER YOUR PURPOSES

What would you like your presentation to achieve? What ideas and viewpoints would you hope that your audience would remember from your presentation? What procedures or changes would you like your audience to enact as a result of your presentation?

Keep in mind that a presentation only gives you a short amount of time (and audience attention) to achieve your purposes. How well you accomplish this will depend on the blend of those three aspects—speaking, visuals, and audience participation. So, identifying your purposes and keeping them before you as you write, design, and practice will help you at each stage of your preparation.

A few tips about purposes:

- Distinguish between purposes for *your topic* and purposes of *ethos*. If, for example, you will be talking about your research, be sure to explain to the audience the goals of your research (as you would in the introduction of an IMRD article). You might make one purpose of your presentation to help the audience understand your research goals and how your project may achieve them. However, a different purpose might relate to the impression of your character (ethos) on the audience. Both types of purposes are important goals for a presentation.
- Achieving your ethos-related purposes will depend on how professional and competent you appear to your audience. These tend to be the aspects of purpose that create the most anxiety for inexperienced presenters.

Doing the exercises in this chapter and paying attention to the suggestions in the earlier sections of this chapter will go a long way toward helping you achieve that professionalism and competence. That is also why we place so much emphasis on learning from experienced presenters, on following the rules and procedures of the venue and event, and why we emphasize your taking advantage of **opportunities to get feedback** on your performance.

- As for research-related purposes, make reachable goals for your presentation. In a 15-minute talk with slides, for example, you will probably not be able to convince your audience of the conclusions of the research project, but you may very well be able to pique the audience's interest in learning more by making your research goals clear and understandable. As described in Section I, you should think of your presentation as a contributor to a research process that will continue well after the presentation and that may very well benefit from the interest you spark in your audience.

CONSIDER YOUR AUDIENCE

Just as you spend preliminary time thinking about the purposes of your presentation, so should you also spend time thinking about your likely audience. This is the same audience-consideration process you go through when writing an article about your research. This presentation is an opportunity for you to **achieve your ethos-related and your research-related purposes**. Don't waste that opportunity by failing to design *this* presentation for *this* audience. That might mean that even if you're talking about the same research on several occasions, you will need to adapt your talk each time.

As you plan each presentation about your research, for example, ask yourself these questions regarding audience:

- What are the likely characteristics of my audience in terms of educational level, interest in my subject, and reasons why they are coming to my talk?
- What can I say about my topic and what visuals can I show that are likely to spark their interest and show my respect for their views and ideas?
- Why is it important that I reach this audience and make them my ally?
- How can I involve them in my presentation in a way that will excite them about my presentation?
- Who can I speak with who knows this audience and can help me answer these questions?

HOW TO SPEND YOUR PREPARATION TIME

1. After you have thoroughly considered constraints, purposes, and audiences (see the above pages), begin to map out your presentation for the allotted time. (If you are part of a presentation team, brainstorm as a group and assign roles.) What topics will the talk and visuals cover? What activities and questions will you provide to involve the audience? How will you divide the time segments? **Note**: This map you create is only a draft. As the presentation develops, you'll adjust the time segments and what happens in each one.

2. Now that you have a rough idea of time allocation, begin to design your talk and visuals. Draft a list of the main points you wish to make and use that list to craft a rough outline of the talk. (Again, as your plans develop, you'll revise this outline.)

3. Begin to fill in the outline with whatever ideas and details you need to get your main points across clearly for this audience and so that it may follow your thinking. How much will you have to define your terms? What examples will you cite? This part of your planning will take considerable time, so be ready to go through multiple drafts as your thinking evolves.

   This third part of your presentation time most resembles writing an article or crafting a poster. However, you'll need to decide if you want to

   - write out the talk as a script, to follow word for word, or
   - craft it as a series of points that you'll use your visuals to speak from more informally.

   What you decide will have a lot to do with how confident you feel speaking on this topic. I've known many presenters who **script** a talk the first time they give it, then become much more spontaneous as their confidence in their knowledge grows until they no longer need a script.

4. Begin to design the slide deck for your talk. (See Section IV, above, for suggestions and also Dos and Don'ts in visual design.) Avoid trying to put your whole talk on slides. Think of the visuals as **concise signs or billboards** that capture an important idea briefly. Some of your visuals will be mostly text, while others will be mostly photos, drawings, flow charts, graphs, or other visual media. One or two may be video clips. Again, you will revise your visual designs many times as your preparation proceeds.

Note: Some presenters *reverse* steps three and four, depending on the types of thinkers they are, whether more verbal or more visual.

5. Practice, practice, get feedback, and practice some more. Practicing aloud is particularly important, because this is an oral performance. If you're planning to record your talk, you can record a few practice runs, which will help you understand both the flow of your presentation and your ethos as a speaker. As you practice aloud for yourself, pay attention to your tempo and imagine yourself as a listener. Even if you're planning for a live presentation, record yourself and play it back. Can you follow what you are saying? Do you sound like a real person in a conversation—or more like one of those voices rattling off side effects and warnings on TV pharma commercials? Chances are you'll find yourself having to cut out some of your talk in order to stay within the time limit.

   Move from one slide to another as you practice your talk. Now imagine yourself as both viewer and listener. Can you follow the words? Can you follow the graphics? Are the visuals remaining before the viewers long enough for them to grasp their meaning? Again, you may find yourself slowing down and having to cut some text in order to keep the attention of viewers and listeners.

   Feedback is especially important for less experienced presenters, but it's useful for anyone. As noted above, try to get constructive comments from experienced presenters or people familiar with the audience and type of venue.

6. Revise any aspect of the presentation accordingly. And practice some more.

7. If possible, before you give the actual presentation, make sure that your visuals (and microphones) work in the presentation venue and with the equipment you will be using. This is a matter not only of professionalism, but also of courtesy to both your audience and fellow presenters. Video clips are particularly susceptible to technical errors, but any facet of visual display and audio communication is subject to malfunctions that can ruin your hard creative work.

Exercise 11F: Interview an experienced presenter in your research field or in a closely related research area. Set up an interview time (for an hour) at this person's convenience. After you have reviewed this chapter, ask a short list of questions that can get this person talking about their presentation experiences—including not only the successes but also the mishaps.

What does this conversation teach you about ways to develop good presentations? Equally important, what does the conversation teach you about building the confidence and skills to prepare and deliver effective presentations that are in fact a pleasure to give?

CONCLUSION

This chapter opened by analyzing the anxiety that many people feel toward giving oral-visual presentations of diverse kinds. It offered many suggestions and tips toward overcoming these fears and producing effective presentations. It looked especially at how presenters can build rapport with audiences—and indeed empower audiences to become allies to the presenter in building knowledge toward better research. The chapter closed with a **seven-step process** for preparing successful, lively presentations that blend the three elements of talk, visual display, and audience involvement.

The next chapter, "Writing Science with Style and Styles," confronts the all-too-common view that STEM writing must be "dry," even boring, if it is to be "good science." But writing that bores is more often a sign of lack of care to communicate clearly and may indeed cover up shoddy method and thinking—which readers are too confused to investigate! This chapter digs into the common stylistic problems of science prose, such as cluttered paragraphing, overuse of passive voice, too much jargon, too many irrelevant numbers, and so on. It offers simple ways that science researchers can make their prose as interesting and effective as their work and ideas.

CHAPTER TWELVE

WRITING SCIENCE WITH STYLE AND STYLES

Topics in this chapter:

I. Keep Sentences Concise with Clear Transitions
II. Guide Your Reader with "Signposts"
III. Use Paragraphs to Emphasize, Not Hide, Your Ideas
IV. Use Numbers to Convince, Not Drown, Your Readers
V. Choose Words to Communicate, Not to Exclude or Intimidate
VI. Revise and Edit to Write with Style

Writing effectively in STEM means caring about **style**. No matter your topic or audience, reaching your reader depends on

- engaging and keeping your reader's **interest**, and
- earning your reader's **respect** for what you have to say.

Keeping interest and earning respect through writing are the results of an appropriate style. The style that is appropriate in any given situation *varies* with the rhetorical questions we have persistently asked from chapter to chapter:

- What are your purposes in this writing?
- What are the expectations of your intended readers?
- How must I tailor my language to the level of familiarity that these readers have with this topic?
- What tone will be appropriate for my purposes and readers?
- What examples can I follow in this genre or medium?

- What elements of design and graphic presentation are appropriate and effective here?

In other words, there is **no single style that represents you in all situations**. A STEM writer's style will not be the same in genres as different as peer-reviewed journal articles, oral-visual presentations, blogs, social media posts, and popular articles in newspapers. For example, geologist Kenneth Hsü's more popular style in his classic book *The Great Dying* (1986) is very different from the IMRD style used in his many peer-reviewed articles. Paleobiologist Hope Jahren's autobiography *Lab Girl* (2016) describes the conventional IMRD style—which has "the precision of a laser scalpel"—that she uses in her peer-reviewed journal papers:

> a rare species of prose capable of distilling ten years of work by five people into six published pages, written in a language that very few people can read and that no one ever speaks. (20)

She contrasts this with the style she uses in her autobiography, a personal style that allows her to explore the emotional challenges and often traumatic events that embody her life as a scientist.

So this chapter is titled "Writing Science with Style and Styles" because the STEM writer needs to learn the **stylistic versatility** that diverse genres, situations, and audiences demand. As we have seen in chapter after chapter, this versatility can only be learned

- through practice,
- through consistent attention to rhetorical questions,
- through close analysis of examples in different genres of STEM writing, and
- through feedback on your work in progress from qualified readers—and possibly from artificial intelligence (AI) programs.

In previous chapters, we've considered the stylistic traits that make up many different genres:

- peer-reviewed IMRD journal articles (Chapters Five and Six),
- peer-reviewed summaries of research (Chapter Seven),
- popular STEM writing in newspapers and magazines (Chapter Eight),
- blogs and social media posts (Chapter Nine),
- posters and infographics (Chapter Ten), and
- oral-visual presentations (Chapter Eleven).

Review those chapters to begin to understand the range of appropriate styles.

However, despite all the differences among appropriate styles, some stylistic traits succeed across genres and situations. These **shared traits** will be the subject of the sections of this chapter.

I. KEEP SENTENCES CONCISE WITH CLEAR TRANSITIONS

STEM professionals are intelligent people. They can connect ideas and understand logical relationships among them. This understanding means that they are able to write single sentences that embody multiple ideas and perform logical connections. This connective skill is essential to the complex thinking needed for the scientific process.

However, from a rhetorical standpoint, complex sentences with multiple connected ideas can pose a problem of comprehension for equally intelligent readers. A sentence structure that works well for you as a writer trying to organize your thinking won't work for readers who are encountering your ideas and connections for the first time.

This difficulty for readers is why we recommend that writers edit their sentences to be concise. When we say that sentences are "concise," we do not necessarily mean that they are short, which is a common interpretation of the term. "Concise" does mean that the sentence does *not* try to incorporate more ideas than a reader can easily comprehend.

How does **editing for conciseness** work in practice?

First, you **draft** your ideas. (And by you, we mean you, not an AI tool—see our discussion of this critical thinking step in Chapters Two and Four.) Do *not* feel constrained to write concise sentences at this early point. Allow yourself to compose ideas and form the connections among them as you feel best represents your thinking. The goal of drafting is to capture your ideas as accurately and fully as you can—and not to worry at this point about comprehension by your intended reader.

Second, as you draft, you'll find yourself **revising** your work as you think of better ways to say what you are thinking—and to spark further thinking. As we explored in Chapter Two, the main purpose of writing at this point in the process is to put your thoughts into words—to perform what we call **writing to think and learn**. Necessarily, as your thinking advances, better ways to express your ideas will come to you—especially as you apply the rhetorical questions to your writing and if you have the help of feedback from trusted readers—and maybe from AI tools.

Third, you will reach a point where you feel that your draft is in sufficiently good shape to **edit** it. Editing is the "fine tuning" of sentences and paragraphs so that you can communicate clearly to your intended readers. Here is where editing for conciseness occurs.

Here's an example. Reread the first paragraph of this section (p. 297), which begins "STEM professionals are intelligent people." When one of us first drafted this paragraph, it came out as one long sentence:

> STEM professionals are intelligent people, one feature of their intelligence being that they are able to connect ideas and understand logical relationships among them, meaning that they possess a connective skill essential to the complex thinking needed for the scientific process, in that they are able to write single sentences that embody multiple ideas and perform logical connections.

Notice that the one long sentence asks you, as a reader, to keep multiple facets in mind as you proceed from clause to clause. You may find yourself rereading parts of the sentence to help you keep all the segments in mind. The sentence structure may also tempt readers to skip over parts of the sentence to get to the end.

In editing for conciseness, we chose to break up the long, overly complex sentence into smaller sentences that allow the reader to grasp one idea before moving on to the next.

Reread the paragraph (p. 297) after editing. You should notice that you can still keep the multiple ideas of the paragraph in mind but without exerting so much energy to do so. Moreover, the more concise sentences have allowed you to remember each of the ideas.

But notice something else. In addition to replacing three commas with periods, we've performed two other actions:

- We've **reordered** part of the long sentence so that the paragraph ends with an idea we want to emphasize: "*This connective skill is essential to the complex thinking needed for the scientific process.*" This idea was lost in the too-complex sentence.
- In two places we added **transitional words** that show the reader the connections between one sentence and the next:

> They can connect ideas and understand logical relationships among them. *This understanding* means that they are able to write single sentences that embody multiple ideas and perform logical connections. *This connective skill* is essential to the complex thinking needed for the scientific process.

By adding the transitional words, we ensure that the reader can see the **flow of ideas** from one sentence to the next. We also enhanced the overall coherence of the paragraph. In the process of editing, look closely at your prose for places where you need clearer transitions—this close editing is one of your most important stylistic acts.

Exercise 12A: In a STEM article of your choice, find sentences that you think could benefit from editing for conciseness.

Choosing one such sentence, subdivide this overly complex sentence into two or more shorter sentences that you feel communicate the ideas more clearly—and with less exertion by your intended reader to understand the connections among ideas.

Pay attention to any need you see to add transitional words from sentence to sentence. Try out some wording.

Also, try to reorder some of the ideas so that the most powerful are emphasized.

What happens when you make these changes? If you are doing this exercise with others, compare the changes you have made and judge the results.

II. GUIDE YOUR READER WITH "SIGNPOSTS"

HEADINGS AND SUBHEADS

In Chapters Five and Six, we commended the writers of the sample articles we studied for their use of **headings** and **subheads**. This tactic is one of the most effective for helping to guide your reader through an article. We call such tactics "signposts" because they help your reader navigate among sections of an article or chapter, when readers look for specific information.

Also, when a reader is scanning an article to see its overall structure and content, headings and subheads can indicate **major topics** in the overall work. For example, you'll note the list of section headings near the start of this chapter. We use these headings to provide signposts of major topics for the chapter as well as to head the appropriate chapter sections.

Headings can vary in specificity. For example, the typical IMRD headings (abstract, introduction, materials and methods, results, and discussion) are generic. They indicate nothing about the *specific* content of the sections in a given article. In contrast, many IMRD articles provide more specific headings that help guide a reader; for example, a discussion section might be subheaded as follows:

- Conclusions
- Limitations of the Study
- Recommendations
- Further Research

Whether a given journal will allow such subheading depends on its editors, but writers and editors often have the flexibility to be more specific. Don't assume a journal's policy. Be guided by (1) your own sense of what headings will be most appropriate for your article and (2) stated policies or examples from your target journals.

For other genres such as posters, websites, and presentations, use your judgment about the specificity of heads and subheads. Keep in mind that readers tend to appreciate more signposts rather than fewer.

Even more specific headings. These are headings that indicate something about the content of the section. Our list of sections for this chapter is an example of headings that indicate content. Another example comes from the December 15, 2017, issue of *Science*, in the article "Superhuman AI for Heads-Up No-Limit Poker: Libratus Beats Top Professionals" (Brown and Sandholm 2017).

Here the authors depart from the standard IMRD practice (except for abstract and Conclusions) to name headings that focus on the article's contents:

- "Game-Solving Approach in Libratus"
- "Abstraction and Equilibrium Finding: Building a Blueprint Strategy"
- "Nested Safe Subgame Solving"
- "Self-Improvement"
- "Experimental Evaluation" (Brown and Sandholm 2017)

Using these more specific, content-focused headings helps your reader navigate the article—and helps to engage the reader's interest.

TEXTUAL MARKERS: BOLDING, BULLETS, AND LINKS

In addition to headings and subheadings, consider using other textual "signposts," such as **boldface** type, **bullet points**, and **links** to other sources. These can be highly useful tools to

- help readers navigate within an article, blog, or website, and
- focus reader attention on points of emphasis.

Unfortunately, these tools are rarely used in print IMRD specialist journals—one of the reasons most peer-reviewed journal articles are hard to plow through, even for other specialists. Online versions of these journals are somewhat more reader-friendly, especially in using links to other online sources. But even these often fail to help their readers navigate, because online versions too often try to emulate old-style print versions.

Fortunately, helpful "signposts" are common in STEM journalism. They are virtually everywhere in well-constructed blogs and websites (see Chapter Nine, p. 245). As you read STEM journalism, blogs, and websites, pay close attention to how the writers use headings, bolding, bullets, and links to attract your attention and guide your reading.

Bolding

As you read STEM publications (print and online), note where and how **boldface type**, as well as *italics* and ALL CAPS, are used to emphasize some words or phrases. Often, bolding and these other tools will only be used in headings. In some publications, they will be used to emphasize words within paragraphs (as we use bolding in this book). Observe yourself as a reader when these tools are used to emphasize specific words or phrases. What happens to your method of reading? More than likely, you will focus on the emphasized words. Then you will read the text that surrounds the bolded words and give it greater attention than you would had some words not been highlighted through bolding.

Bolding is especially useful in longer paragraphs—to keep readers attentive as they move from sentence to sentence. (But see Section III, below, on better paragraphing.)

Bullet Points

This tool is a common organizational device in PowerPoint, but it can be highly useful in print and online articles of *all* kinds. Bullet points help to **organize related items**:

- Lists of ingredients, components, steps in a process, etc.
- Related explanations for a phenomenon

Visually, bulleted lists say to a reader, "These items are related and should be organized in your mind that way." Also, bulleting will draw a reader's eye to the list in a way that normal paragraphing does not. Try this: write a list of items in a sentence, with a comma between each pair of items. Then reformat the list as bullet points. You'll notice at least two things: (1) the bulleted items stand out in a way that the items in a sentence do not; (2) the bulleted list requires more space. When you use bulleting to draw your reader's eye, you have to make a choice regarding what's more important to you: saving space *or* holding attention? If making sure that your reader pays attention is worth the loss of space, use bullets.

Notice also that bullets work to draw attention because of 3 factors:

- The bullet **point** itself is a visual mark.
- The **indenting** of the bulleted item draws the reader's attention.
- The **white space** that surrounds the list focuses the reader on the list items.

In other words, the bulleted list works because it looks different from the paragraphs around it.

This difference means that you need to use bulleted lists sparingly. If you use them too often, say more than twice per page, they lose that appearance of difference. This is why PowerPoint slides that consist of one set of bullets after another can become boring. It's also why singers, speakers, and TV advertisers have to modulate their tones between soft and loud (and some silence!) so that listeners will maintain attention. Think of bulleted lists as a way of making your tone "louder" amid paragraphs.

Links

Do you usually think of links on a webpage only as a navigational device that takes you off the page to somewhere else? But links are also very visual markers that draw attention. Again, observe yourself as you scan a webpage. How do links appear? As highlighted words? As photographs? Do you find yourself looking for links? If so, then you know the power of links to grab your attention.

In peer-reviewed journals online, links are now commonly used

- to take you to **supplements**, usually of materials and methods and of results, and
- to take you to **sources** in the list of references.

However, they are now becoming more commonly used to **highlight terms** within paragraphs. When used in this way, the link draws the reader's eye in the same way that bolding does—but even more powerfully, because the link offers the promise of further information beyond the page.

But *caution*: as with bolding and bullets, links won't work to hold attention if they become too commonplace. We all know webpages that are masses of links. These are great—if the web designer wants you to use the webpage as an **index**, such as a list of references with links to the articles. But if you as a writer/designer want to focus your reader's attention, **use**

links sparingly. Remember: every time your reader moves off your page to follow a link, chances are greater that the reader won't return to your page. Instead of using a link to highlight a term, consider using bolding, which will draw attention but keep the reader on your page. Links to further readings can always be included in the reference list at the end of your article.

Exercise 12B: Browse a number of STEM articles and websites to see how

- bolding
- bullet points
- links, and
- headings

are used. Be sure to choose at least one piece of STEM journalism, one IMRD article in a peer-reviewed journal, and one STEM website or blog for your analysis.

For bolding, bullet points, and links: How is each of these tools used in each type of document? List these uses. How do they enhance navigation of the article or site? How do they attract your attention?

For headings: Do the headings work to help you navigate the site or article? Do the headings give you any information about the content of the site or article? Do the headings help to engage your interest?

If others are doing this exercise with you, compare the results each of you observes.

Exercise 12C: Using the data from Exercise 12B, evaluate the effectiveness of the tools used in each sample document you have selected.

Do a separate analysis of each tool used (bolding, bullet points, links, headings). How has the tool use been effective? How could it be improved?

Next, write your overall evaluation of the uses of these tools across the samples you've studied:

- Which sample has used the tools most effectively for navigation and for attracting reader attention? How and why?
- Which sample could use the most improvement? How and why?

Lastly, how might you apply the results of this analysis to your own use of these tools in your writing and design?

If you are doing this exercise with others, compare findings and conclusions.

III. USE PARAGRAPHS TO EMPHASIZE, NOT HIDE, YOUR IDEAS

Think of a paragraph *not* as a collection of sentences in a group—but as a **visual arrangement** of words and sentences in a white space. Thinking of a paragraph this way goes against what you may have been taught in school about paragraphing, but thinking about the paragraph as a visual marker is much more useful as you think about the design of your writing to reach readers.

What constitutes a "paragraph"? As you've gone through school, you may have heard different definitions of the term and seen different "rules" for constructing paragraphs. We've seen the paragraph defined as "a complete thought" or "a group of sentences that make up a single idea." A commonly heard rule of paragraphing is that "a paragraph may *not* have only one sentence." Another rule is "paragraphs should have at least three sentences" (or four, five, or six, depending on the person who made up the so-called rule).

However, if you observe contemporary constructs of the paragraph in print and online STEM documents, you'll find a wide range—and rarely does the number of sentences in the paragraph matter.

Instead, what mostly matters to contemporary document designers is the visual effect of the paragraph. Will the paragraph draw the reader's attention? Will it keep the reader engaged? Will it use space economically without hiding important ideas in the middle?

For example, in STEM journalism (Chapter Eight), paragraphs tend to be short—one-sentence paragraphs are frequent. (Look back in Chapter One, p. 23, at the *New York Times* article on endangered Australian birds for a good example.) There are good practical reasons for one-sentence paragraphs. Print newspapers have multiple columns (four or more) across a single page, which means that paragraphs must be narrow. Limiting the number of sentences to one or two allows paragraphs to be broken every two inches or so. Quoted material (for example, from interviews) usually gets its own one-sentence paragraph.

Remember: short paragraphs get more attention.

When these print documents are translated into an online format, many articles appear to be collections of one-sentence paragraphs. While this arrangement might appear "choppy" to some readers, most readers are accustomed to the design and have no problem following the train of thought.

Similarly, popular STEM magazine journalism (Chapter Eight) also uses multiple columns per page, most often two. Plus, the frequent use of photographs in STEM popular magazines often further narrows the area that can be used for text. As a result, paragraphs in these journals rarely exceed three sentences, and often include fewer.

Likewise, posters and infographics (see Chapter Ten) privilege shorter paragraphs, because they, too, give considerable space to photos, tables, graphs, and other visual elements that restrict text.

Presentation slides (Chapter Eleven) emphasize short paragraphs for a different restriction: time. Listeners to oral-visual presentations can only process a small amount of text before the speaker must move on. Large blocks of text take time to read and limit absorption of content.

Where the long paragraph still rules. In contrast, the IMRD peer-reviewed journal article usually maintains the idea of the long paragraph, sometimes extending to much of a page. Full-length books by STEM authors, as well as STEM-relevant articles in such magazines as *The New Yorker*, also hold true to the older convention of the long paragraph. Because of this purely conventional practice, readers must work harder to locate and remember important ideas (see Chapter Six, p. 154, for more on the negative effects of this holding to tradition).

Peer-reviewed IMRD journals are gradually adapting to the new multimodal, more flexible electronic paradigm (see Chapters Three, Five, and Six), but the old print conventions still reign in many.

What does the contemporary range of paragraph lengths mean? In practical terms, what this range of designs means for writers is that there are no simple rules. As elsewhere in the topics we've covered, writers need to make paragraphing choices based on

- the type of publication they want to strive for,
- reaching their intended readers, and
- achieving their purposes for the writing.

More importantly, this range of practices means that writers should feel free to adjust paragraph length according to what they want to emphasize in their writing—and not according to some arbitrary "rule" based on number of sentences or on merely conventional practice.

We recommend using the visual power of the short paragraph to focus your reader's attention on important ideas.

So, if you've drafted an article with long paragraphs, review your draft to see if you've buried an important idea in the middle of a long paragraph. If so, break up that long paragraph into two or more paragraphs so that the key ideas you wish to emphasize get the attention they deserve.

If you use AI tools in revising your drafts, please note: AI paragraphs often default to the tradition of the long paragraph. Don't let this AI practice make you a boring writer! Take what AI has produced and make your writing more interesting and effective by editing it for paragraphing and signposts.

Exercise 12D: The Distance Test. To judge how paragraph length affects reader attention and focus, try this test:

- Using a print STEM source of your choice, hold a page of the sample far enough from you so that you cannot read the text but still see the shape of the paragraphs. Where does your eye go to on the page? If you are like most viewers, your eye will go to the shortest amount of text surrounded by the most white space. Also, if you are like most viewers, your eye will avoid a dense block of text—a long paragraph.
- Further, note that your eye also goes to other visual markers—such as bolding, bullets, and headings—that stand out from the mass of gray text. These markers, along with short paragraphing, work to emphasize certain words.

Now try the Distance Test with one of your own writings. Does your eye go to words and sentences that you consider important? Or does it go to less important words and sentences? Rearrange your paragraph design so that your most important ideas get the attention you want them to.

IV. USE NUMBERS TO CONVINCE, NOT DROWN, YOUR READERS

Way back in Chapter One (p. 27), we compared the use of numbers in a peer-reviewed IMRD article from *Biological Conservation* with the more sparing and focused use of statistics in a *New York Times* article based on the STEM IMRD article. We returned to this topic of statistical reporting in STEM writing in Chapters Three, Ten, and Eleven—and most importantly in Chapter Six (pp. 151–74) on the results sections of peer-reviewed IMRD articles.

Here we reinforce a point made in those chapters:

- Use of statistical data should be both **accurate** and **selective** in STEM writing. Writers should resist the impulse to inundate (aka "drown") readers with data points and multi-variable graphics that more often confuse and intimidate than communicate.

The hard-to-resist impulse to overwhelm readers with numbers comes from two sources.

First, younger scholars in particular often feel that they must demonstrate how hard and thoroughly they've crunched the numbers in order to establish their credentials—their rhetorical ethos. They often fear that, if they do not report all their data, they will be accused of superficial analysis—or of choosing only the data points that fit their hypothesis.

Therefore, they fill their results sections with densely packed tables and multi-factor graphs that merely bog down their readers in unnecessary data points that confuse their readers and that hide the data they should emphasize. This is *not* the way to build respect for your research.

Remember that results sections in IMRD articles are meant to identify the most important data from a study and argue for the significance of these critical data (see Chapter Six for review).

If you feel that you must display fuller data in your article, do so in a well-organized **supplementary section** which you can link to your results section. Indeed, many online peer-reviewed journals now require such supplements. (If you are writing for a print publication, include the bulk of your data in an **appendix**, or list the URL of the online supplement in the list of references.)

Second, many inexperienced STEM writers are fearful of their responsibility to select the critical data and to argue for their significance. Hence, they merely dump their poorly analyzed data on their readers and hope that significance will magically appear. Again, review Chapter Six for suggestions on how to construct a successful argument in an IMRD context. Chapter Five will also be helpful, as it shows writers how to construct the IMRD article so that it tells a coherent, convincing story about the research.

NUMBERS IN STEM JOURNALISM, POSTERS, AND PRESENTATIONS

When writing outside the IMRD context, always use statistics sparingly. Decide which statistics

- will be most convincing to a non-specialist audience,
- will most accurately represent the impact of your research, and
- will be most easily understood by your audience.

Keep in mind that numbers—if used sparingly and if carefully chosen—can be very powerful in your arguments. In addition, a key statistic will draw your reader's eye as strongly as other visual markers.

We highly recommend your reading such popular STEM magazines as *Scientific American*, *National Geographic*, *National Wildlife*, and *Catalyst* so that you might observe how professional STEM journalists and excellent scientists/writers carefully select the most telling statistics and present them so that they attract reader attention and make arguments convincing.

Similarly, review Chapter Ten (Posters and Infographics) and Chapter Eleven (Oral-Visual Presentations) for suggestions on displaying statistical information in genres that are greatly restricted in available space (posters and infographics) or time (presentations).

Exercise 12E: Choose an IMRD article in your research field and analyze it for its use and presentation of statistical evidence, particularly in its results section. Note the following:

- Is statistical information divided between (1) key statistics reported in the text of results and (2) much more statistical data reported in a data supplement?
- In the results section, do statistics appear carefully chosen for ease of understanding and for their significance to the argument of the article?
- Do you at any point bog down in tables of confusing data or in trying to interpret multi-variable charts and graphs?

How would you suggest improving the selection and presentation of statistics for effectiveness and clarity? Conversely, where does the article present statistics clearly and convincingly?

V. CHOOSE WORDS TO COMMUNICATE, NOT TO EXCLUDE OR INTIMIDATE

When our undergraduate students read academic research articles, one aspect of much of this writing that stands out to them is the great amount of **unfamiliar language**. They have trouble getting through the dense undergrowth of sophisticated vocabulary and highly technical terminology (insider jargon)—as well as unexplained acronyms. To them, the articles are obviously addressed to a readership to which they do not feel they belong.

As a result, the research literature that should be reaching undergraduate STEM majors fails in its mission and winds up making them feel excluded from the scientific community and afraid to venture further into its thorny linguistic territory.

If that is not bad enough, another "lesson" that this murky academic prose appears to teach students is that "success" in academic writing comes from using this highly specialized vocabulary. Far too often, teachers see student writing that tries to impress with big words that students seem not to understand—but that they feel they must use in order to seem intelligent.

Not only are undergraduate students put off by the intimidating language, but even graduate students and other fellow professionals have difficulty trudging through many articles. In a recent meeting with STEM PhD students, one of us mentioned that we were writing this text, and they exclaimed how impenetrable are so many of the articles *in their own research field*. So intimidated are these STEM researchers that they feel embarrassed to admit their difficulty in reading and understanding. Even less likely are they to call out their published colleagues for their unintelligible writing.

In the next chapter, we will focus on a few key ways to edit your sentences so that your writing is easier to follow and understand. Here we focus on one of those editing issues: **word choice**.

> **Recommendation**: In editing your STEM writing, think first and foremost about your intended reader's understanding of the words you choose.

If you have not systematically edited for word choice before, we recommend doing so **word by word**, with particular emphasis on terms that you feel are most important for your intended readers to understand.

Here's a sample paragraph from a classic article, with a few such terms highlighted:

> Let me mention another proof of the bivalency of beryllium which may have passed unnoticed, as it was published in the Russian chemical literature. Having remarked (in 1884) that the density of such solutions of chlorides of metals, MCl_n, as contain 200 mols. of water (or a large and constant amount of water) regularly increases as the molecular weight of the dissolved salt increases, I proposed to one of our young chemists, J. Burdakoff, that he should investigate the beryllium chloride. If its molecule is $BeCl_2$ its weight must be = 80; and in

> such a case it must be heavier than the molecule of KCl = 74.5, and lighter than that of MgCl = 93. On the contrary, if beryllium chloride is a trichloride, BCl_3 = 120, its molecule must be heavier than that of $CaCl_2$ = 111, and lighter than that of $MnCl_2$ = 126. Experiment has shown the correctness of the former formula, the solution $BeCl_2$ + 200 H_2O having (at 15°/4°) a density of 1.0138, this being a higher density than that of the solution KCl + 200 H_2O (= 1.0121), and lower than that of $MgCl_2$ + 200 H_2O (= 1.0203). The bivalency of beryllium was thus confirmed in the case both of the dissolved and the vaporised chloride. (Mendeleev [1897] 1999, 410)

Marking words in this way helps to focus your attention on both

- the technical **accuracy** of the wording, and
- readers who are likely to **understand—or, conversely, miss**—your intended meanings.

For example, three key terms in this paragraph from Mendeleev are "bivalency," "beryllium," and "density." The author (and the English translator) assume the reader's understanding of these terms, because they are not defined. However, the reader who does not understand these critical terms is likely to miss the point of the experiment described.

If you as a writer are attuned by practice to thinking of your intended reader's level of technical knowledge, then you can decide *if and how* you might change wording—or if you need to **explain in simpler language** terms that you feel are important to your article.

Again, this is an exercise to be performed in the editing phase of your writing process, not in drafting. Thinking so intently word by word too early in your process can severely cramp your ability to conceive of and express ideas. The "writer's block" so often encountered by writers of all disciplines and genres is usually the result of a person's so strictly questioning each and every word in an early draft that nothing gets composed.

But once you get to that editing stage, you may find that while you are focusing on reviewing word choice, you may also discover that there are other changes you wish to make in order to clarify your thinking for your reader. Any stage of the writing process, even editing, can lead you to insights that will improve your writing as a whole.

Note: If you are using AI tools to revise or edit an early draft, by all means do *not* assume that the AI version is

- accurate in its content,
- in any way meant to be accessible to your intended readers, or
- going to attract their attention.

It is your job as a rhetorically savvy writer to revise and edit an AI-produced draft to really meet the needs of your readers.

Exercise 12F: Choose a favorite IMRD article in your research field and re-read it from two different points of view:

- First, assume that you are an intended reader of this article. As you read, mark several words that you are not sure you understand as the writer uses them in the text. Focusing on each word, write one or more words or short phrases that you think could substitute for the actual word—or that could explain the term in a way that you would be sure to understand. (You may need to research the term so that you can understand it sufficiently before you try any substitute wording. By doing this exercise, you are in effect editing the piece, and in the way that most readers "edit" their reading as they read.)
- Second, imagine that you will be paraphrasing (see Chapter Four, pp. 102–03) this article for a reader who does not have technical familiarity with its language, but who might be interested in its topic and conclusions. Try to think of an actual person who might be this reader. Mark several key terms in the article that this reader would not understand. Then write substitute words or phrases that you might choose to try to explain each of these terms to this reader. (By doing this exercise, you are beginning to do what STEM journalists do in translating IMRD journal prose for non-specialist readers.)

If you are doing this exercise with others, compare your substitute wording. What do you learn from one another?

VI. REVISE AND EDIT TO WRITE WITH STYLE

You may have noticed that all of the techniques we've explained in this chapter so far for improving the style of your STEM writing are more specifically meant to improve the clarity and navigability of your writing. These techniques are also meant to help you emphasize what you regard as the most important ideas in your writing. Together,

- clarity,
- navigability, and
- emphasis

are vital traits of "writing with style."

This final section of the chapter on "style and styles" advises STEM writers on how to use the stages of the writing process to achieve these three goals.

USING THE IDEA OF A "WRITING PROCESS"

Most of the chapters in this book go into detail—and supply exercises—to help you become a better writer by teaching you to become a better reader of all kinds of STEM documents. Equally essential are the many concrete techniques in Chapter Two for "getting started" with your writing by using techniques for "writing to learn." Your reading and your writing-to-learn prepare you for drafting a specific project—and then for getting good "feedback" from fellow writers to help you revise your drafts. The current chapter (and the chapter to follow) offer many techniques for that stage of the writing process known as editing—which can't occur until you have a revised draft of your project with which you are reasonably satisfied.

When we talk about these varied learning tasks,

- using reading to improve writing,
- writing to learn,
- drafting,
- getting feedback,
- revising, and
- editing,

we are talking about typical parts of a multi-stage "writing process" that *every* writer engages in, but that differs from writer to writer.

Each phase of a writer's process contributes to the whole, and each is necessary. You can try to short-circuit your process, usually because you think it will save time; but if you are a conscientious writer, you'll usually find yourself going back to add steps you thought you could do without.

For example, most writers will try from time to time to do just one draft of an assignment ("I don't have the time to do more! It's due tomorrow!"), but unless the writer has done many similar writings before, the results of the "one draft method" will be disappointing, and the writer will look for opportunities to revise and edit later on—or to do the reading and note-taking and other preliminary planning work that should have been done first.

Particularly important for editing for style is all that preliminary work. While you might begin a project with the goal in mind of reaching a specific type of reader—perhaps a non-STEM readership or, conversely, a small group of highly specialized researchers—you can't fine tune your prose until you have a revised draft that you are reasonably sure has expressed your ideas accurately.

Following this step-by-step method is particularly critical if you have used AI tools at any stage of your process, as the note (above) explains.

Only if you have carefully followed the steps of the process can you look closely at your revised draft and carry out the specific actions that this chapter has described:

- Keep sentences concise with clear transitions
- Guide your reader with "signposts"
- Use paragraphs to emphasize, not hide, your ideas
- Use numbers to convince, not drown, your readers
- Choose words to communicate, not to exclude or intimidate.

A (humorous?) note of caution. Be aware that even when you are fine tuning your prose through editing, you may—and often will—discover that what you thought was a perfect piece of explanation or argument isn't as perfect as you'd imagined. As you reread, don't be surprised if you realize ways to rephrase, or reorganize, or even completely rewrite a section of your work. The writing process is insidious in that way. The technical term is "recursive"—meaning that the stages of the process may repeat at any time. So, even while you are editing, for example, the number of sentences in one paragraph, you may see that that paragraph is not as thorough or accurate as you'd imagined, and that you need to go back to your notes (or the lab) so you can revise that section. Ah, the joys of writing!

Exercise 12G: Find a piece of your own STEM-related writing that you consider finished—perhaps a published article, a completed grant application, an old course paper, last month's report to your boss, your poster for last month's conference, etc.

Edit it using the five operations described in this chapter (and listed just above). Review the sections of the chapter if you need to in order to perform the operations effectively.

What changes do you make? What do you discover about the clarity, navigability, and emphasis in your document?

Do you discover other changes you'd like to make? On the other hand, what do you think you did well in that document?

CONCLUSION

Writing with "Style" in STEM is about writing to interest your reader and keep that reader interested—while also earning that reader's respect for the accuracy and insightfulness of your research. We've focused on **five stylistic operations** in this chapter that are meant to

- reduce clutter, so meanings are **clear**,
- help your readers **navigate** smoothly through your prose,
- and **emphasize** the ideas you want highlighted.

We've ended by outlining the common stages in the writing process—especially how each stage of that process can result in writing that crisply and clearly achieves what you set out for it to do.

Chapter Thirteen, "Editing Sentences," focuses even more sharply on the key building block of all STEM writing: the sentence. It offers tips for editing grammar, voice, length, tone, and word choice—all to make meanings stand out.

Note that we distinguish *editing* from the crucial work of *revision*, by which writers modify their ideas and presentation throughout the process of writing to make it more effective.

CHAPTER THIRTEEN

EDITING SENTENCES

Topics in this chapter:

I. Why We Must Edit
II. Cut Unneeded Words
III. To "We" or Not to "We"
IV. Action vs. Passivity—Tuning Your Voice
V. Punctuate to Accentuate

While a scientist takes great care to keep a worksite clear, clean, and ready for work, sentences in science prose often can't do their work because meaning is hidden by clutter of all kinds. This chapter offers tips for **editing grammar, voice, length, and tone** to make sentence meanings stand out.

Be sure to read Chapter Twelve, "Writing Science with Style and Styles," for other tips on editing for clarity, navigation, and emphasis throughout the documents you are writing. Chapter Twelve also features the valuable final sub-section—"Using the Idea of a 'Writing Process'"—on how to use *all* stages of your writing process to help you edit at the right time.

Here we look at the basic building block of your prose: the sentence.

I. WHY WE MUST EDIT

No one drafts edited prose.

Inexperienced writers imagine that their own halting, tortured, confusing drafts show what poor writers they are. They imagine that if they were good writers, they wouldn't need to worry endlessly about their sentences, changing words and word order, then changing the sentences back to what they were at first, then trying out new words, until they hit their deadlines and just turn in whatever they have at that point. (Or much worse: giving the task of writing to a chatbot, and then handing in that plagiarized document as their work!)

Or they rely on "hope for the best." In absolute dread of writing, they wait until the deadline is almost upon them, then manage to scramble out a draft of the report or the proposal or whatever, turn it in, and hope for the best.

Either way, writing for them is no fun—largely because they suffer under the delusion that they should just be able to "think" the right words onto the screen. If they are scientists, that attitude toward writing makes about as much sense as thinking that designing and carrying out a perfect test or experiment happens quickly and in one try.

Even for experienced writers and researchers, the process—whether of writing or of test design—requires creative thought, trying out ideas, looking at results, getting some help, then making adjustments, seeing more results, and so on. But the big difference between the novice and the veteran is *attitude*. The veteran knows that the process is painstaking, and so takes it in stride: builds in the time for the process, doesn't panic—and often really enjoys the tinkering, and the surprises, and the talking it over with colleagues, and finally getting it almost, ever so close, to what they come to see as success.

Think of editing sentences as the final adjustments you'd make to the article or proposal or presentation you've worked so painstakingly on: like the final adjustments to the test design/apparatus *after* all the preliminary thinking and mock-ups and pilot tests and feedback and redesign. You couldn't do it earlier because you didn't know if you had anything worth presenting. But now you do, and so you concentrate on putting your best foot forward so that your audience will understand clearly and emphatically what you were trying to show. You care about this final stage, because you don't want your audience to say "What?" or, worse, "So what?" And you'd like them to feel at least a little bit of your excitement.

That's what editing is—and must be.

II. CUT UNNEEDED WORDS

As you edit, look systematically for **words you can cut from your sentences without losing meaning**. When we draft, we typically use more words than we will eventually need. Why? Because in drafting we try out words that we think we might need to get our meaning across. When we edit, we already have a good idea of which words will be most impactful, so we can remove what we no longer need. A text pared down to the words that carry clearly the most meaning will be more effective for your readers.

PLACES TO LOOK FOR CUTS

"Hedge" Words—Cautious Qualifiers

These are the words that we throw into sentences when we want to be cautious in what we claim:

> probably
> likely
> perhaps
> usually
> often
> sometimes
> presumably
> mostly
> approximately

Note that all of these words estimate quantities or frequencies, but are not specific. We call them "hedges" because writers use them to try to protect ourselves from the criticism that they don't have the specific results to make a definite claim. When you see them in your revised draft for a STEM audience, cut them, and substitute more precise evidence. For example,

> **Revised Draft**: "In most cases, the index registered between 46 and 83 in intensity (78%)."
>
> **Edited**: "In 78% of cases, the index registered between 46 and 83 in intensity."
>
> **Change**: we've replaced the "hedge" term "in most cases" with the precise figure, 78%.

Conversely, if you see a hedge word in your revised draft, but do not have precise wording (or stats) to substitute in that spot, ask yourself if you can with confidence just remove the hedge word. Are you being overly cautious by using the hedge? Did you in fact just put it in because it is conventional to do so? If so, **cut the hedges**. See if you are comfortable with the statement you are making.

> **Revised Draft** (with highlighted terms): "In most cases, the index registered between 46 and 83 in intensity (78%). This probably indicates a wider range than most predictors will be comfortable with.

> However, a narrower range, between 56 and 70, accounts for only 10% of measurements, possibly too few for predictors to use as a viable range for susceptibility."
>
> **Edited**: "In 78% of cases, the index registered between 46 and 83 in intensity. This indicates a wider range than predictors will be comfortable with. However, a narrower range, between 56 and 70, accounts for only 10% of measurements, too few for predictors to use as a viable range for susceptibility."

If you are *not* comfortable with deleting these hedges, your need will not be for a hedge word, but for precise pieces of evidence that will justify your claim. Hedges have their place in writing for some audiences, but here we are focused on STEM audiences.

Repetition of Wording

Along with hedges, look for **repetitions of wording** as targets for cutting. STEM writers use repetition to ensure that their readers can clearly see the connection between topics in one sentence (or paragraph) and another. In Chapter Twelve (pp. 297–99), we noted that it is important for writers to **show transitions** between sentences, so that readers do not get lost. These transitions require some repetition.

However, too much repetition is not purposeful for showing transitions. For example, in the following abstract, we've inserted a repetition (highlighted) that a writer might choose to cut as unneeded for reader understanding:

> Biological inorganic carbon fixation proceeds through a number of fundamentally different autotrophic pathways that are defined by specific key enzymatic reactions. Detection in (meta)genomes of the enzymatic genes thought to be needed for these pathways is widely used to estimate the contribution of individual organisms or communities to primary production. Here we show that the sulfur-reducing anaerobic deltaproteobacterium *Desulfurella acetivorans* is capable of both acetate oxidation and autotrophic carbon fixation, with the tricarbolyxic acid cycle operating either in the oxidative or reductive direction, respectively. (Mall et al. 2018; highlight added)

Your choice of keeping or deleting repetition will depend on your judgment of what an intended reader **needs to know** in order to make sense of each sentence. So, look for repetition in your own prose, and ask yourself, "Do I need this repetition for my reader's understanding, or can I cut it?"

"Filler" Phrases

Your (and our) draft prose tends to contain groups of words that we use almost unconsciously as "filler" between ideas in a sentence. We are so accustomed to using these phrases that we may not even be aware of doing so. Even in editing, they will escape our detection. It may only be when we face a strict word limit, as in a proposal or application, that suddenly we look closely to see if there are "filler phrases" we can cut.

In the first sentence of the sample cited above, we've highlighted two phrases that could qualify as filler—if the writer judges them as unnecessary for reader understanding:

> Biological inorganic carbon fixation proceeds through a number of fundamentally different autotrophic pathways that are defined by specific key enzymatic reactions. (Mall et al. 2018; highlight added)

Try reading the sentence with and without the highlighted phrases. Do the phrases make a difference in your understanding?

Exercise 13A: Choose one piece of your own STEM writing and one piece of published STEM writing by a different author(s). Perform targeted editing that focuses on

- hedge words,
- repetitions of wording, and
- filler phrases.

What do you discover? What seem to be your favorite hedge words and filler phrases?

What changes did you make in your own prose? What changes would you suggest the author(s) of the other article make?

If you are doing this exercise with others, compare your edits.

III. TO "WE" OR NOT TO "WE"

Our STEM students have always said that they've received advice, often conflicting, about when and whether to use the first person ("I" and "we") in their writing. Most commonly, they've heard, "Never use the first person in science writing." Why? "Because your writing is about the science; it's not about you."

Conversely, they've also heard, and with much observation to back it up, that more and more IMRD journals have relaxed this stricture. It is now common for peer-reviewed IMRD publications to allow limited use of "we," as in the following example (highlighted) from the same abstract excerpted above (p. 318) from *Science*:

> Here we show that the sulfur-reducing anaerobic deltaproteobacterium *Desulfurella acetivorans* is capable of both acetate oxidation and autotrophic carbon fixation.... (Mall et al. 2018; highlight added)

"We" is also used in descriptions of methods, as in the following from the same article:

> To further elucidate the fate of citrate, we incubated cell abstracts of *D. acetivorans* under anaerobic conditions....
>
> ... [W]e also observed six 13 (superscript)C NMR multiplets....
>
> Using liquid chromatography/mass spectrometry (LC/MS), we found that the CoA/acetyl-CoA ratio.... (Mall et al. 2018; highlight added)

Note that in all these instances, the use of "we" does not deflect the reader's attention from the materials studied—it does not make the article "about" the research team. Instead, **it accurately places the team inside the research procedures**. The use of "we" helps the reader to visualize more precisely and efficiently the procedures used. Trying to avoid the use of "we" by shifting to passive voice only creates an artificial and inaccurate—and possibly confusing—picture of the research scene. Note the difference:

> **Use of "we"**: To further elucidate the fate of citrate, we incubated cell abstracts of *D. acetivorans* under anaerobic conditions.... (Mall et al. 2018; highlight added)
>
> **Avoiding "we" by shifting to passive voice**: To further elucidate the fate of citrate, cell abstracts of *D. acetivorans* were incubated under anaerobic conditions.... (Mall et al. 2018; highlight added)

The first version is clear, direct, and easy to visualize by a reader. The second is not.

WHAT ABOUT "I"?

STEM writers increasingly use "we" in IMRD journals in the ways we noted. But **"I" is almost never used**—for the simple reason that almost all published peer-reviewed research is co-authored. So using or not using "I" is irrelevant in the IMRD literature.

WHERE TO AVOID USING "WE"

In the IMRD peer-reviewed literature, avoid beginning sentences with expressions such as

- "we think,"
- "we believe," or
- "we feel."

Such phrases do push reader attention away from your materials and methods and place it on your state of mind or your emotions. Such phrases also betray your lack of confidence in your methods and results.

If you find yourself wanting to use "we think" or "we feel," try writing the sentence without those two words. Does the idea stand by itself? If so, then great. If not, then adding those two words won't help matters. **Look instead for wording that focuses on the evidence** that backs up your claim. Indeed, a beginning phrase as simple as "Our results show ..." will place the emphasis where it belongs. And if your evidence is as strong as you claim, then your confidence is justified.

"I" AND "WE" IN STEM JOURNALISM

As described in Chapters Eight and Nine, STEM journalism, books and biographical essays by scientists, and blogs give writers stylistic freedom that the IMRD research literature does not allow. This stylistic freedom includes acceptance by readers—even the expectation—that STEM writers will use the first person in their writing.

Using the first person in these genres goes beyond the mere use of "I" and "we." STEM writers are free to write about their experiences, their backgrounds, and their emotional reactions to history and public issues. One of the great pleasures of reading these genres is that they allow us to see the researchers *as people* that the research literature keeps appropriately hidden.

Indeed, the acceptance of science by a broad public depends on a balance between

- the **rigorously empirical methods and style** of the IMRD peer-reviewed literature, and

- the **multi-sided, human, and vulnerable portraits** of scientists that come through in the other genres.

Exercise 13B: Study one or more IMRD articles in your research area. Note in particular if and how the writers use "we." What purposes seem to be served by these uses (e.g., to describe methods or note observations)?

Do you find any uses of "we" that seem inappropriate? If so, in what way?

Conversely, do you find any opportunities for clarity or precision that are missed because the writers did not use "we" but instead used a passive construction? Edit the wording, and see how the meaning becomes more precise.

Compare your findings with those of others doing this exercise.

Exercise 13C: After doing Exercise 13B, study two or three STEM journalism articles relevant to your research area. Note how the first person ("I" and "we") is used in these articles.

Describe which purposes of the articles are served by these uses:

- experiences of the researchers?
- attitudes toward issues or challenges?
- description of methods?
- perspectives on results?
- other?

Do you feel that how the first person is used in these articles helps to make researchers and their research understandable to and relatable by a wider audience? Explain.

Compare your findings with those of others doing this exercise.

IV. ACTION VS. PASSIVITY—TUNING YOUR VOICE

As noted in the previous section, our students have all heard strongly worded advice about using the first person ("I" and "we") in their STEM writing. In contrast, they have heard only vague suggestions, if any, about using either **active or passive voice** in their sentences.

Nevertheless, whenever STEM writers go out of their way to avoid using "we" in describing a procedure—for example,

> "We carried out three separate trials..." *becomes*...
> "Three separate trials were carried out..."

—those writers are changing a clear, more active sentence into one that is less direct and less easy to visualize. Rhetoricians call the "were carried out" option a **passive construction** because it erases the doer of the action from the scene—and it calls on the reader to *assume* who might have conducted the trials. (Did the research team do the work? We *assume* so, but the article in fact does not say.)

Fortunately for readers of science, STEM journalists mostly avoid passive constructions, because they know that **active voice** attracts and keeps readers.

Watch how active voice (highlighted) functions in this excerpt from the *Los Angeles Times*, by Amina Khan (2018):

> Could the TRAPPIST-1 star system host a life-friendly planet? Planetary scientists probing the bevy of small, Earth-sized worlds surrounding this not-too-distant star have found that at least a few may not look too different from our own.
>
> The findings, described in a study in Nature Astronomy and a paper in Astronomy and Astrophysics, highlight the planetary system as a promising target in the search for life on other worlds.
>
> Seven Earth-sized planets circle TRAPPIST-1, an ultracool dwarf star that lies about 40 light-years away. With about 9% of our sun's mass, this star is very small and dim. That means that even though TRAPPIST-1's planets lie so close to their star that they would sit within Mercury's orbit around the sun, a number of them still could be the right temperature to hold water. And because they're so close to their star, the planets circle it frequently, making them easier to study.

We've highlighted the action verbs in the first three paragraphs of the *LA Times* article. Notice that these simple verbs describe either physical actions ("circle," "hold," "probing") or states of rest ("lie," "sit") that are easy for readers to visualize. Notice also that the **subject words** that carry out these actions ("star system," "scientists," "planets") come before the verbs—the sentence position easiest for readers to understand.

But readers are fortunate in another way.

Active voice also dominates the IMRD article on which this *LA Times* piece is based. The following excerpt comes from the abstract of "Atmospheric Reconnaissance of the Habitable-Zone Earth-Sized Planets Orbiting TRAPPIST-1" (de Wit et al. 2018) in *Nature Astronomy*:

> Here, we report observations for the four planets within or near the system's habitable zone, the circumstellar region where liquid water could exist on a planetary surface (12–14). These planets do not exhibit prominent spectroscopic signatures at near-infrared wavelengths either, which rules out cloud-free hydrogen-dominated atmospheres for TRAPPIST-1 d, e and f, with significance of 8σ, 6σ and 4σ, respectively. Such an atmosphere is instead not excluded for planet g. As high-altitude clouds and hazes are not expected in hydrogen-dominated atmospheres around planets with such insolation (15,16), these observations further support their terrestrial and potentially habitable nature.

We've highlighted in this paragraph four pairs of subjects and verbs that the writers have put in active voice, the word order that readers most readily understand. For example, the first pair

> we report

concisely moves the reader on to what the team reports, followed by its several types of significance. The final pair

> these observations further support

moves the reader directly to the main conclusion about the observations—that these planets may be potentially Earth-like.

By using **active constructions**, the writers help their specialist audience achieve and keep focus on the main ideas.

It is especially important to observe that using active voice in the IMRD research article does *not* mean that writers shift focus away from "the science" and on to themselves. A sentence that begins "we report observations" is no less empirical than one that begins passively with "observations are reported. . . ." In fact, the first version is *more* empirical because it names a doer of the action of reporting, which the second version neglects to state.

Further, note that only one of the four highlighted instances of active voice in the excerpt uses "we." For example, the sentence that begins

> These planets do not exhibit prominent spectroscopic signatures at near-infrared wavelengths ...

uses active voice. Watch what would happen to the sentence if we were to change active to passive:

> Prominent spectroscopic signatures at near-infrared wavelengths are not exhibited by these planets...

Note how cumbersome and hard to follow we've made the sentence. By putting the long phrase "Prominen... wavelengths" at the beginning, we force the reader to memorize a complex passage before the reader even learns the subject ("these planets") or verb ("exhibit") of the sentence. **When editing, STEM writers need to prioritize their readers' ease of understanding**. If they do, they will use far more active than passive voice.

Exercise 13D: Change the following sentences from passive voice to active:

1. The seven samples were observed over one week and daily measurements compared.
2. The above characteristics of maximum contaminant levels (MCLs), with particular emphasis on those of hexavalent chromium, were shown to predominate in March 2022 samples from only five of the wells.
3. The results reported in the earlier analysis (2,4) have been corroborated by more recent studies (8,12) carried out in temperatures below -5C.

Do this exercise with others, if possible. Compare the changes you've made. There may be more than one effective version. Keep in mind that the goal of editing is to make meaning as clear and easy for your readers as you can. The goal is not to make passive or active voice an end in itself.

Exercise 13E: Review a piece of your own STEM prose—if possible, one that you had considered finished.

Pay special attention to active and passive voice. Try editing the text to increase the amount of active voice. Note wording and word order changes you have to make to effect this change in voice.

What challenges do you encounter? What new vocabulary do you find yourself having to use?

Do you have to deal with your own resistance to using active voice in STEM? Or is this an easy transition for you?

This exercise might be an incentive for you to read more IMRD articles that effectively use active voice. Looking for these articles, and studying them, will increase your versatility as a writer.

V. PUNCTUATE TO ACCENTUATE

When you edit your sentences, don't ignore punctuation. **Punctuation** is another tool that can help you make your ideas clear and understandable to your readers.

Most of us think of punctuation as a requirement of correct grammar, and yes, it is surely that. But punctuation is also a feature of the **visual presentation** of your writing. In Chapter Twelve, we emphasized your creating "signposts" to help guide your reader (pp. 299–303). As your reader follows each sentence, the punctuation you use, such as the commas that enclose this phrase, also serve as signposts to help your reader understand how to read this sentence.

But more than that: some punctuation, such as the colon (:) that we used after "that" in this sentence, proclaims a particular message to your reader. The colon says loudly, "I promise important information. Pay attention!"

In this section, we note the visual power of several marks of punctuation and suggest how you might use them to enhance your STEM writing.

THE COLON

The **colon** is often used to announce a numbered or bulleted list. Together, these markers create a strong visual effect. See, for example, the bulleted list of "purposes" in the middle of Exercise 13C, above, p. 322.

BULLETS

See Chapter Twelve (pp. 301–02). In conventional twentieth-century grammar guides, **bullet points** were not recognized as marks of punctuation since they were rarely used earlier in that century. But their now ubiquitous use in PowerPoint and their prevalence in websites and in business and advertising have made the bullet important in written communication of all kinds. But note my caution in Chapter Twelve about the risks of overuse.

THE DASH

Known as "double hyphens" in British usage, the **dash**—when used carefully—will draw attention to words you want to emphasize. Visually, the dashes that surround the chosen words allow them to stand out. Note how in the previous sentence, the words "when used carefully" draw your eye because of the dashes that separate them from the rest of the sentence.

Dashes should be used when you want them to create sharp focus on particular words. Do *not* use them as you would parentheses or commas, just to separate out small bits of explanation, as in the following:

Parentheses: The results reported in the earlier analysis (2,4) have been corroborated by more recent studies (8,12).

Commas: The first test, March 23 at Cameron Creek, was less successful than the second, April 14 at Peel Run.

Dashes: The fourth test—by far the most successful—took place June 18th before the entire committee at Arroyo San Miguel.

An amazing thing about dashes is that they can enclose **any grammatical construction** and still be correct. Note the following examples:

- The fourth test—which the entire committee observed—brought spectacular results.
- The fourth test—LeClerc called it "the day that made our careers"—began in cold, rainy weather.
- The fourth test—does anyone who was there not recall it vividly?—almost didn't take place because DuMaurier and Medvedev grumbled about the weather.

EXCLAMATION POINTS

Ironically, STEM journalists rarely use them to show emphasis, even though that is the purpose of the exclamation point. In informal communication, such as Facebook posts, brief emails, and tweets, they are still ubiquitous. In these very short forms, the **exclamation point** serves the purpose of showing the writer's excitement.

But the exclamation point is at best a blunt instrument of emotion—experienced writers have many means at their disposal to draw their readers' attention and engage their ongoing interest. Further, the emergence of **emojis** in electronic communication to show a range of emotions has to a great extent replaced the simple exclamation point even in short forms.

The fate of the exclamation point exemplifies what can happen through overuse. Seeing an exclamation point rarely excites a reader, unless it appears rarely, and only if the reader's emotion is actually generated by the words that the mark is supposed to emphasize. Too often, the overused exclamation point has the opposite effect: the reader suspects that the writer may be trying too hard to kindle excitement about something that doesn't deserve it, or may even be showing false emotion.

Exercise 13F: Choose one or more IMRD research articles in your research area. Read the article(s) closely for uses of punctuation for their visual impact. More than likely, you will find many, many uses of commas and periods, some uses of parentheses and colons, and far fewer of dashes and bullets. Most of the time, unfortunately, you will not find that writers have used punctuation to enhance visual appeal.

Your main objective here, nevertheless, is to judge the visual design of the article(s), and to look for opportunities to enhance reader attention and comprehension through better design. Be sure to pay specific attention to uses of punctuation that improve visual impact.

Choose several sentences that could be made more visually effective by using dashes and colons with bulleted lists. Try out some changes and judge the results.

If you are doing this exercise with others, compare your changes and their effects.

QUESTION MARKS

Amid paragraphs of explanation in the sections of the IMRD article, a **well-placed question** can make your reader take notice. For example,

> What would happen if...?
>
> If we raised the temperature one degree per hour, when would...?
>
> Was this an anomaly, or was it part of a pattern?

As in conversation, when a speaker's question challenges a listener, a question in writing **changes the relationship** between the writer and the reader. A question calls a reader to action—to think, to wonder, to engage. The reader can no longer just sit back passively and wait for all to come from the writer. The reader is now under pressure.

Notice the difference to you as a reader between this sentence *about* a question (1) and a question itself (2):

1. "The relation between perception and action remains a fundamental question for neuroscience." (Assaneo and Poeppel 2018)
2. What is the neural relationship between perception and action?

STEM writers miss easy opportunities to engage their readers if they fail to edit sentences like (1) into actual questions, like (2).

Note how the authors effectively use a question later in the introduction to the same article:

> Is coupling between acoustic stimuli and auditory cortical activity linked to the sensorimotor machinery? (Assaneo and Poeppel 2018)

Certainly, stating actual questions can be overused, as can any stylistic strategy. But the *occasional* shift from exposition to question can be the **emotional change of pace** that readers need to remain engaged in an article. As you edit your sentences, look for those opportunities.

And by all means, **use some key questions in your oral-visual presentations.** A slide that would be yet another column of bullet points turns into an actual dialogue between you and your audience when you challenge them with a question.

Exercise 13G: Choose one or more IMRD research articles in your research area. Note if and where questions (with questions marks) are used by the writers.

If so, do you find that the questions are well chosen to gain reader attention and challenge reader thinking?

If not, can you see opportunities in the article(s) where questions might be used effectively?

Further, can you see places where a question might be indented to give it even greater attention?

If you are doing this exercise with others, compare your responses.

Hint: Some typical places for questions would be in the abstract, the introduction, and the discussion, though materials and methods and results also offer opportunities.

SEMICOLONS VS. PERIODS: VISUAL IMPACT

The **period** is by far the most common mark used to indicate the end of a sentence. **Semicolons** may also be used for this function, when the writer wants to **show a relationship** between two complete sentences. (Colons and dashes may also serve this function.) For example, here is a statement from the same neuroscience article we cited just above:

> Interareal synchrony might support sensorimotor integration; however, the key properties have not been characterized, in particular the sensitivity to syllable rate, arguably the most fundamental property of speech perception and production. (Assaneo and Poeppel 2018)

As you read this sentence quickly, notice how your pace does *not* slow as you encounter the semicolon at the end of the first sentence. Readers tend to move past **weak punctuation** such as a semicolon in order to get to the next part of the sentence. But watch what happens to your reading speed when you end the sentence with a period after "integration."

> Interareal synchrony might support sensorimotor integration. However, the key properties have not been characterized, in particular the sensitivity to syllable rate, arguably the most fundamental property of speech perception and production.

The period slows you down and allows you to digest the topic sentence. Then the new sentence, beginning with the strong conjunction "however," shows you the relationship with the prior idea.

Keep in mind that either way of punctuating this statement is grammatically correct. Each part that we have identified is a complete sentence, so the two parts may be separated by a period, or by a semicolon. From the standpoint of visual impact, however, the period is superior, for the reason described above.

Caution: You may have heard at some point in your schooling that "you can't begin a sentence with a conjunction." This is another one of those "advice" statements that have been inflicted on writers (often children) as so-called rules. As a result, writers have used semicolons much more often than they have needed to in instances such as these:

- The solution evaporated in the hot lab; therefore, we had to make more and keep it refrigerated.
- The Ptolemaic diagrams portrayed orbits as circles; but telescopic observation showed them to be elliptical.

Again, either the period or the semicolon is grammatically correct. Judge visual impact to help you decide which to use in a given sentence.

Exercise 13H: Choose one or more IMRD research articles in your research area. Edit the article(s) for appropriate uses of periods or semicolons as sentence boundaries. Be particularly aware of the visual impact of the sentences.

More than likely, you will find many more uses of the period than of the semicolon. When you do find sentences that are linked with semicolons, could you improve reader comprehension by removing the semicolons and inserting periods? Try out the different arrangements and judge the effects.

However, even if you do *not* find much use of the semicolon, you may find many instances where sentences strike you as needlessly long—where many ideas are jammed together in single sentences that might be separated into shorter sentences that would help readers comprehend the multiple ideas (see Chapter Twelve, p. 299, and Exercise 12A for more on this problem and how to handle it).

Edit such sentences and judge the results. If you are doing this exercise with others, compare your findings and your suggested edits with those they have made.

Exercise 13I: Using artificial intelligence (AI) tools, generate an answer to a STEM question of interest to you. Then, within that chatbot answer, try out some or all of the suggestions from this chapter for making your sentences more alive, uncluttered, and clear. Judge the results.

Doing this exercise with others can reveal some clever possibilities for making chatbot prose more interesting. And you'll also have ideas for your own original STEM writing.

CONCLUSION

Editing is the final—and always necessary—step in the writing process. In this chapter, we have looked at multiple ways to trim the fat from sentences, plus ways to make your writing sharper and more vivid for your readers.

You should never be satisfied with dull prose when you have so many tools to bring your writing to life.

REFERENCE LIST

Akashi, Muhammad, and Noam Soker. 2017. "Shaping Planetary Nebulae with Jets in Inclined Triple Stellar Systems." *Monthly Notices of the Royal Astronomical Society* 469 (3): 3296–306. https://doi.org/10.1093/mnras/stx1058.

Akil, Huda, and Eric J. Nestler. 2023. "The Neurobiology of Stress: Vulnerability, Resilience, and Major Depression." *PNAS* 120 (49): e2312662120. https://doi.org/10.1073/pnas.2312662120.

Anthes, Emily, and Chang W. Lee. 2024. "Should We Change Species to Save Them?" *New York Times*, April 14. https://www.nytimes.com/2024/04/14/science/australia-wildlife-assisted-evolution.html.

Aristotle. 2006. *On Rhetoric: A Theory of Civic Discourse.* 2nd ed. Translated by George Kennedy. London: Oxford UP.

Assaneo, M. Florencia, and David Poeppel. 2018. "The Coupling between Auditory and Motor Cortices Is Rate-Restricted: Evidence for an Intrinsic Speech-Motor Rhythm." *Science Advances* 4 (2). https://doi.org/10.1126/sciadv.aao3842.

Beyrer, Chris. 2021. "The Long History of mRNA Vaccines." Johns Hopkins Bloomberg School of Public Health, October 6. https://publichealth.jhu.edu/2021/the-long-history-of-mrna-vaccines.

Bhattachargee, Yudhijit. 2024. "The New Science of Stress." *National Geographic*, June.

Broad, William J. 2017. "A Giant Nuclear Blast, but a Hydrogen Bomb? Too Soon to Say." *New York Times*, September 3. https://www. nytimes.com/2017/09/03/science/north-korea-bomb-test.html.

Brown, Noam, and Tuomas Sandholm. 2017. "Superhuman AI for Heads-Up No-Limit Poker: Libratus Beats Top Professionals." *Science*, December 17. https://doi.org/10.1126/science.aao1733.

Cohen, Phoebe A., Justin V. Strauss, Alan D. Rooney, Mukul Sharma, and Nicholas Tosca. 2017. "Controlled Hydroxyapatite Biomineralization in an ~810 Million-Year-Old Unicellular Eukaryote." *Science Advances* 3 (6). https://doi.org/10.1126/sciadv.1700095.

de Wit, Julien, Hannah R. Wakeford, Nicole K. Lewis, et al. 2018. "Atmospheric Reconnaissance of the Habitable-Zone Earth-Sized Planets Orbiting TRAPPIST-1." *Nature Astronomy* 2 (February): 214–19. https://www.nature.com/articles/s41550-017-0374-z.

Durso, Andrew. 2017. "How Many Snakes Are Venomous and How Many Are Constrictors?" *Life Is Short, but Snakes Are Long* (blog). August 31, 2017. https://snakesarelong.blogspot.com/2017/08/.

Einstein, Albert. (1916) 2014. *Relativity: The Special and General Theory*. Reprint, CreateSpace Independent Publishing Platform.

Hawkins, Ed. 2018. "Warming Stripes." *Climate Lab Book* (blog), May 22. https://www.climate-lab-book.ac.uk/2018/warming-stripes/.

Hsü, Kenneth. 1986. *The Great Dying*. Harcourt Brace Jovanovich.

Innis, Harold A. (1950) 2022. *Empire and Communications*. University of Toronto Press.

Jahren, Hope. 2016. *Lab Girl*. Knopf.

Khan, Amina. 2018. "TRAPPIST-1 Planets Are Rocky and Have Complex Atmospheres, New Studies Show." *Los Angeles Times*, February 6. https://www.latimes.com/science/sciencenow/la-sci-sn-trappist-planets-habitable-20180205-story.html.

Kizilcec, René F., Andrew J. Saltarelli, Justin Reich, and Geoffrey L. Cohen. 2017. "Closing Global Achievement Gaps in MOOCs." *Science* 355 (6322): 251–52. https://doi.org/10.1126/science.aag2063.

Kobak, Dmitry, Rita González-Márquez, Emőke-Ágnes Horvát, and Jan Lause. 2025. "Delving into LLM-Assisted Writing in Biomedical Publications through Excess Vocabulary." *Science Advances* 11 (27). https://doi.org/10.1126/sciadv.adt3813.

Larson, Daniel, and Tom Misteli. 2017. "The Genome—Seeing It Clearly Now." *Science* 357 (6349): 354–55. https://doi.org/10.1126/science.aao1893.

Ledford, Heidi. 2016. "CRISPR: Gene Editing Is Just the Beginning." *Nature* 531 (7593). https://www.nature.com/articles/531156a.

Lewis, Ricki. 2017. "Brain Cancer in Kids: Tailoring Treatments Based on Mutations." *DNA Science* (blog). October 5, 2017. https://blogs.plos.org/dnascience/2017/10/05/brain-cancer-in-kids-tailoring-treatment-based-on-mutations/.

Little, Crispin T.S. 2017. "Life at the Bottom: The Prolific Afterlife of Whales." *Scientific American*, May. https://www.scientific american.com/article/life-at-the-bottom-the-prolific-afterlife-of-whales/.

Mall, Achim, Jessica Sobotta, Claudia Huber, et al. 2018. "Reversibility of Citrate Synthase Allows Autotrophic Growth of a Thermophilic Bacterium." *Science* 359 (6375): 563–67. https://doi.org/10.1126/science.aao2410.

Malone, J. Patrick. 2009. "Self Recognition and the Rise of What Most of Us Refer to as Personhood." YouTube, 4:24. Accessed 2019. https://www.youtube.com/watch?v=pNqirOJ5qAw.

Mehlenbacher, A.R. 2019a. "Registered Reports: Genre Evolution and the Research Article." *Written Communication* 36 (1): 38–67. https://doi.org/10.1177/0741088318804534.

Mehlenbacher, A.R. 2019b. *Science Communication Online: Engaging Experts and Publics on the Internet*. Ohio State University Press. https://library.oapen.org/handle/20.500.12657/24833.

Mehlenbacher, A.R., and C.R. Miller. 2017. "Intersections: Scientific and Parascientific Communication on the Internet." In *Landmark Essays on Rhetoric of Science: Case Studies*, 2nd ed., edited by Randy Allen Harris. Routledge.

Mendeleev, Dmitri. (1897) 1999. "The Periodic Law of the Chemical Elements," translated by George Kamensky. In *19th Century Science: An Anthology*, edited by A.S. Weber. Broadview Press.

Motluk, Alison. 2008. "Mirror Test Shows Magpies Aren't So Bird-Brained." *New Scientist*, August 19. https://www.newscientist.com/article/dn14552-mirror-test-shows-magpies-arent-so-bird-brained/.

NASA. n.d. *A Guide to Climate Change for Kids*. Climate Kids. Accessed 15 October 2024. https://climatekids.nasa.gov/kids-guide-to-climate-change/.

NASA. 2024. "Annual Arctic Sea Ice Minimum Area 1979–2024, with Graph." NASA/Goddard Space Flight Center Scientific Visualization Studio, October 4. https://svs.gsfc.nasa.gov/5395/.

Nelson-McDermott, Catherine, Don LePan, and Laura Buzzard, eds. 2014. *Science and Society: An Anthology for Readers and Writers*. Broadview Press.

Ou, Horng D., Sébastien Phan, Thomas J. Deerinck, Andrea Thor, Mark H. Ellisman, and Clodagh C. O'Shea. 2017. "ChromEMT: Visualizing 3D Chromatin Structure and Compaction in Interphase and Mitotic Cells." *Science* 357 (6349). https://doi.org/10.1126/ science.aag0025.

Pavlova, Alexandra, Peter Selwood, Katherine A. Harrisson, et al. 2014. "Integrating Phylogeography and Morphometrics to Assess Conservation Merits and Inform Conservation Strategies for an Endangered Subspecies of a Common Bird Species." *Biological Conservation* 174 (June): 136–46. https://doi.org/10.1016/j.biocon.2014.04.005.

Pavlova, Alexandra, Sara Petrovic, Katherine A. Harrisson, et al. 2023. "Benefits of Genetic Rescue of a Critically Endangered Subspecies from Another Subspecies Outweigh Risks: Results of Captive Breeding Trials." *Biological Conservation* 284 (August): 110203. https://doi.org/10.1016/j.biocon.2023.110203.

Plato. 1871. *Gorgias.* Translated by Benjamin Jowett. Independently published.

PLOS. n.d. "About PLOS BLOGS." PLOS BLOGS Network. Accessed January 2019. https://blogs.plos.org/about/.

Prior, Helmut, Ariane Schwarz, and Onur Güntürkün. 2008. "Mirror-Induced Behavior in the Magpie (*Pica pica*): Evidence of Self-Recognition." *PLOS Biology* 6 (8). https://doi.org/10.1371/journal.pbio.0060202.

Roig, Miguel. 2015. *Avoiding Plagiarism, Self-Plagiarism, and Other Questionable Writing Practices: A Guide to Ethical Writing.* Office of Research Integrity, US Department of Health and Human Services. https://ori.hhs.gov/sites/default/files/plagiarism.pdf.

Russell, John. 2015. "Drug Ads Include a Lot of Warnings—Probably Too Many, FDA Say." *Chicago Tribune*, September 4. http:// www.chicagotribune.com/business/ct-confusing-drug-ads-0906-biz-20150904-story.html.

Safecast. 2025. "Map of Radiation Readings." https://map.safecast.org/?y=37.191&x=140.175&z=9&l=0&m=0.

Sahu, Kailash C., Jay Anderson, Stefano Casertano, et al. 2017. "Relativistic Deflection of Background Starlight Measures the Mass of a Nearby White Dwarf Star." *Science* 356 (6342): 1046–50. https://doi.org/10.1126/science.aal2879.

Sanchez, Maria. 2016. "Undergraduate Helps Tackle Global Threat of Tick-Borne Disease." *One Health* (blog). UCDavis, September 2. https://www. ucdavis.edu/one-health/undergraduate-helps-tackle-global-threat.

Sanchez-Bayo, Francisco, and Koichi Goka. 2014. "Pesticide Residues and Bees—A Risk Assessment." *PLOS One* 9 (4): e94482. https://doi.org/10.1371/journal.pone.0094482.

Science Sam. 2023. "Next time you look in the mirror, remember this microscopic reminder." TikTok, December 8. https://www.tiktok.com/@science.sam/video/7310331265553616134?lang=en.

Shulman, Seth. 2017. "Standing Strong for Science and Democracy." *Catalyst*, Winter. https://www.ucsusa.org/publications/catalyst/standing-strong-for-science-and-democracy#.XBbCIydCdE4.

Soler, Manuel, Tomás Pérez-Contreras, and Juan Manuel Peralta-Sánchez. 2014. "Mirror-Mark Tests Performed on Jackdaws Reveal Potential Methodological Problems in the Use of Stickers in Avian Mark-Test Studies." *PLOS One* 9 (1). https://doi.org/10.1371/journal.pone.0086193.

Weber, A.S, ed. 1999. *19th Century Science: An Anthology*. Broadview Press.

CREDITS

Broad, William J. "A Giant Nuclear Blast, but a Hydrogen Bomb? Too Soon to Say," *The New York Times*, 3 September 2017. Copyright © 2017 The New York Times. All rights reserved. Used by permission and protected by the Copyright Laws of the United States. The printing, copying, redistribution, or retransmission of this Content without express written permission is prohibited.

Chevallier, Olivier, Connor Black, Julia Balog, Sara Stead, Steven Pringle, Zoltan Takats, Christopher T. Elliott. "Real Time Detection of Fish Fraud Using Rapid Evaporative Ionisation Mass Spectrometry (REIMS)," presented at the 2016 Metabolomics conference, Dublin. Reprinted with permission.

Ledford, Heidi. "CRISPR: Gene Editing Is Just the Beginning," *Nature* 531 (2016): 156–59. Copyright © 2016. https://doi.org/10.1038/531156a. Reprinted by permission of Springer Nature, conveyed through Copyright Clearance Center, Inc.

Snook, Ryan. Illustration from "CRISPR: Gene Editing Is Just the Beginning," *Nature* 531 (2016): 156. Reprinted with permission.

IMAGE CREDITS

"Annual temperatures in Toronto from 1841–2017," by Ed Hawkins, *Climate Lab Book*, 22 May 2018. Used under CC BY-SA 4.0 license: https://creativecommons.org/licenses/by-sa/4.0/.

Chemical construction of Glyphosate, Wikimedia Commons. Credit: Yikrazuul, https://commons.wikimedia.org/wiki/ File:Glyphosate.svg.

Glyphosate 3D images, Wikimedia Commons. Credit: Benjah-bmm27, https://commons.wikimedia.org/wiki/File:Glyphosate-3D-balls.png; https://commons.wikimedia.org/wiki/File:Glyphosate-3D-vdW.png.

Drive Map screen capture from Safecast.org. Used under CC BY-NC 4.0 license: https://creativecommons.org/licenses/by-nc/4.0/.

From "Climate Kids: A Guide to Climate Change for Kids." Courtesy NASA/JPL-Caltech.

NASA Scientific Visualization Studio object screen capture. Courtesy of NASA/Goddard Space Flight Center Scientific Visualization Studio. The Blue Marble data is courtesy of Reto Stockli (NASA/GSFC).

INDEX

About the Publisher

The word "broadview" expresses a good deal of the philosophy behind our company. Our focus is very much on the humanities and social sciences—especially literature, writing, and philosophy—but within these fields we are open to a broad range of academic approaches and political viewpoints. We strive in particular to produce high-quality, pedagogically useful books for higher education classrooms—anthologies, editions, sourcebooks, surveys of particular academic fields and sub-fields, and also course texts for subjects such as composition, business communication, and critical thinking. We welcome the perspectives of authors from marginalized and underrepresented groups, and we have a strong commitment to the environment. We publish English-language works and translations from many parts of the world, and our books are available world-wide; we also publish a select list of titles with a specifically Canadian emphasis.

broadview press